Patrick Moore's Practical Astro

Other Titles in this Series

The Urban Astronomer's Guide

A Walking Tour of the Cosmos for City Sky Watchers

Rod Mollise

Springer

Rod Mollise
Mobile, AL, USA

British Library Cataloguing in Publication Data
A catalogue record for this book is available from the British Library

Library of Congress Control Number: 2005932858

Patrick Moore's Practical Astronomy Series ISSN 1617-7185
eISBN: 1-84628-217-9
ISBN-10: 1-84628-216-0
ISBN-13: 978-1-84628-216-4

Printed in Singapore/KYO

9 8 7 6 5 4 3 2 1

Springer Science+Business Media
springer.com

Contents

Introduction

The Urban Astronomer's Guide is the result of a crazy idea for a book that came to me nearly twenty years ago. I knew deep sky objects—nebulae, star clusters, and galaxies—could be seen from the city. Plenty of amateur astronomers were braving countless streetlights in search of distant marvels. But there was little written information available on viewing the deep sky from urban and heavily light-polluted suburban locations. The only mention of the subject I found in astronomy books and magazines was the stern warning that it was impossible to get a good look at anything other than the Moon and planets from urban locales. I knew this wasn't true—I'd observed the entire Messier list from my bright backyard and had a lot of fun doing it. I was sure many more amateur astronomers "trapped" amid city lights would also love to see deep sky wonders from home—if only they had a little information and encouragement. Solution? The book you hold in your hands.

I spent many, many hours observing the objects that form the sky tours included in this guide. But that was the enjoyable part of the project. The hard work was done by my friend and fellow observer, Pat Rochford, and by my dear wife, Dorothy. They didn't share much of the observing fun; instead they devoted themselves to the tasks of checking the manuscript, and, most importantly, providing the encouragement I needed to keep going with what some people told me was an "impossible" concept for a book. Thanks are also due—in spades—to John Watson. Just when I was ready to toss this idea aside, he, like Pat and Dorothy, kept me on track. Dorothy, Pat, and John—this one is for you.

Rod Mollise
Selma Street
Summer, 2005

PART I

Telescopes and Techniques

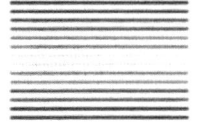

CHAPTER ONE

The Whys and Hows of Urban Observing

Have you gone out into your backyard or garden and looked up at the night sky lately? If you're a seasoned amateur astronomer living in the city, especially an astronomer interested in the deep sky, the universe of objects beyond the Solar System, you probably haven't. The conventional wisdom is that the quarry deep sky observers hunt—star clusters, galaxies, and nebulae—doesn't show up well, or at all, from the typical sodium-streetlight-pink urban sky. Every veteran city-bound amateur probably spent some time observing from home as a novice when every night was an adventure and not a single clear evening was to be wasted. But with experience and a growing orientation toward deep space, most city dwelling astronomers eventually desert the backyard for occasional trips to darker locales—an astronomy club "dark site," a friend's vacation home or farm, or an organized star party.

Trips to dark sites are great, but wouldn't you like to get out with your wonderful telescope more often? That's what this book is all about. Whether you're a novice amateur or a deep sky veteran, it will show you how to enjoy *night after night* of wonderful sights from the comfort of home. There's an amazing amount to be seen, even under the brightest skies. What I am going to do is take you on a walking tour of the cosmos. We'll travel from depressing city lights to the wonders of deep space. You'll learn what you need to pack for these hikes, what's to be seen out there, and how best to see it. The bulk of the book consists of ready-made seasonal tours of the heavens, but you'll also learn how to plan your own night sky journeys. Before considering the "how" of urban amateur astronomy, though, let's talk more about the "why."

Yes, observing from perfect country skies is wonderful, but an emphasis on dark site observing comes at a price for the urban-dwelling amateur: if you rely only on these opportunities, you'll usually wind-up observing once a month—if that frequently. "Once a month" is a far cry from the "every clear night" of novice days, but for today's amateur that's often as good as it gets. Organized club star parties are usually confined

3

to weekends closest to the new Moon, and, while an individual with a personal dark site can theoretically get out deep sky observing more often than that, the facts of modern life—two career families and long workdays—tend to rein things back to once-a-month. Distance is another complication that cuts the frequency of observing runs for the urban astronomer. Getting to dark skies means driving 40–60 miles from the center of even a medium sized city. If you have to travel an hour or two, set up the scope, and then allow time for tearing things down, packing and returning home, you are not going to be doing much weeknight observing. This once a month syndrome (which may be reduced to "every couple of months" due to poor weather) means that the urban deep sky observer is usually badly out of practice.

Being lost in space is a feeling well known to the city-based astronomer. It's been a couple of months since you were last out observing at the club observatory, and, even then, you didn't see much since the New Moon came on a partly cloudy evening. Tonight is different. You're at the Texas Star Party, your yearly getaway under the superbly bright stars and dark skies of the Southwest U.S. desert. Not a light in sight. Velvety blackness and stars everywhere. And there you stand, not quite feeling in harmony with the cosmos. The telescope that was so easy to assemble in your active novice days now seems slightly puzzling. Where do you insert the bolts that attach the tripod to the mount? What was that quick-and-accurate polar alignment method that once seemed so easy? Naturally, the constellations, with their scads of stars visible in dark skies, look a little unfamiliar, but getting oriented would be easier if you at least remembered which *bright* star was which. If that weren't bad enough, when the telescope is finally assembled and aligned, objects that once looked spectacular don't seem to show as much detail as they did in the past. You almost feel as if you've forgotten how to observe. You have.

Sir William Herschel, arguably the greatest amateur astronomer of all time and a professional musician, often likened observing with a telescope to playing a musical instrument, and was of the opinion that astronomical observing, like music, requires constant practice. If you've experienced the above lost in space feeling, you know Sir William was right. Observing is a complex series of tasks, from gathering equipment for the evening's run, to developing a list of observing targets, to getting the best view of a galaxy. Without constant repetition these skills grow rusty. How good are you at any complicated task you only perform once every month or two?

What's the answer? It would be nice if we all enjoyed dark skies from home, but light pollution is not going away tomorrow. Many dedicated amateur and professional astronomers are working to reduce this curse of modern times, but it's unlikely that the average urban amateur's skies are going to get better any time soon. The answer for the city observer is simple and lies close at hand: despite bright skies, observe every clear night. From the backyard, the rooftop, the secure park, the science museum parking lot, or any place in the city where there's an open view of the sky from safe surroundings.

What Can You See from the City?

"Well, that's OK for the Solar System boys. They can do well downtown. You don't need dark skies to view Jupiter, Saturn, or the Moon, but I don't care about that stuff.

I want to observe the deep sky, and you *have* to have a dark site to see *anything* beyond a few of the very brightest objects." Wrong. There's a *lot* to be seen by the patient, educated deep-sky observer from almost any urban site, including:

- The entire Messier list, even M74, M33, M76, and M97, the supposed "hard ones."
- Many NGC objects, and not just open star clusters, though you can feast on as many of those as you desire.
- Supernovae burning in the hearts of distant galaxies.
- The beauty of the classical constellations in their stately march across the sky as the seasons change.
- The comings and goings of those intergalactic tramps, the comets.
- Hordes of asteroids tracing their lonely paths through the Solar System.
- The animals that form our urban ecosystem and survive unnoticed under the foot of Man.
- The looks of wonder on the faces of family, neighbors, and friends as you show them sky marvels from the friendly surroundings of home.

Come join me on a typical city evening's observing adventure. Tonight, my instrument of choice is my "big" scope, an inexpensive Meade 12.5-inch Dobsonian reflector. Depending on my goals, I might have chosen an 80-mm short-tube refractor, a Celestron Nexstar 11 Schmidt Cassegrain telescope (SCT), or just a pair of binoculars. On this evening the 12.5-inch scope is appropriate because, in addition to observing some Messier galaxies, I'll be searching for a supernova, an aged and obese star that's ending its life in a spectacular explosion near the heart of a distant galaxy. The 12.5-inch telescope provides generous aperture for supernova hunting, and it is also surprisingly easy to set up. I carry its "rocker-box" mount outside, plunk it down, manhandle the tube out the back door and onto the rocker box and I'm ready to go.

With the scope assembled, I turn to the evening's observing list. Some fellow amateurs find it amusing that I go to the trouble of drawing up detailed lists and charts for an informal peek at the sky from my backyard. But, in truth, detailed planning is probably more important for the urban astronomer than for those blessed with dark skies. If the sky is clear, country astronomers can turn their scopes to any quarter of the heavens and be rewarded. We urban observers have to be more discriminating. Before setting the scope up, I had a look at the virtual sky with the aid of *Skytools 2*, a computerized astronomy planning and charting program that runs on my PC. I used it to help me select interesting objects nearing the meridian for my date and time— objects as high in the sky as they'd ever get for my location. This selection process allows me to escape some of the worst effects of light pollution near the horizons.

As the sky darkens and enfolds me and my beloved telescope, I begin to get into the rhythm of sky and land. Sure, if I concentrate I can hear traffic on the busy thoroughfare just two blocks away, but my mind filters that out. I only hear the comforting chirp of crickets, smell the spring smells of garden greenery on the warm air, and see the inviting glimmer of the first stars to grace the evening sky. The sky itself? Oh, it's not pristine. Far from it. The horizons are ringed by the gaudy pink of countless sodium arc streetlights. Even on a crisp winter night the short exposure of Orion in Plate 1 is obviously fogged due to heavy sky glow. Conditions are even worse tonight in the hazy atmosphere of spring. But the great constellation Leo is riding high tonight. His sickle, hindquarters, and even a few of his dimmer stars sparkle into view as the day ends. These are not the skies of a southwestern desert, never have been, never will be, but they are still beautiful in their own right.

Soaking-up ambience is nice, but I am hungry for the deep sky. Referring to my charts and using the large-aperture finder mounted on my Dobsonian, I "star-hop" to my first target, galaxy Ml05. After just a little hunting around—I'm very familiar with the area, since my backyard site allows me to get out every clear spring night and tour Leo and Virgo—I have M105, an elliptical galaxy, in the field of my eyepiece. Once I have it centered I increase my magnification a little to provide a pleasing view and just stand and look for a few minutes. Before I began urban observing, I would have doubted that M105 would even be visible from the city. But there it is. It is not only visible but "bright" and displaying as much form and substance as any elliptical galaxy can, a bright core surrounded by an extended, circular envelope of nebulosity. There's more. As I continue to stare at the field, two more galaxies pop into view. Little NGC smudges, companions to bright M105.

After spending an hour hopping from galaxy to galaxy across the Lion, I remember my "special object" for the evening, a supernova that has appeared in galaxy NGC 3877. Not knowing quite what to expect, I move the scope to the location in Ursa Major where this nondescript spiral lurks. I've never looked at it before, but my big finder and the wide-field eyepiece in the main scope help me pin it down without too much trouble, despite the fact that it lies in an area of my sky that is almost completely barren of stars to the naked eye. The galaxy is not much to look at (and probably wouldn't be much even under dark skies), but it's detectable in a medium-power eyepiece.

And there's the supernova, a fiery speck close to NGC 3877's center that gives the galaxy a seeming "double nucleus." For a moment I'm a little awestruck. This isn't the first supernova I've seen, but the thought of the significance of the photons pouring into my eye from this ancient, violent event never fails to evoke wonder. Well I know that I probably wouldn't have seen it at all if I'd disdained the backyard. By the time I could have organized a trip to a dark site—a couple of weeks, probably— the supernova would likely have dimmed past the point where I could detect it from the darkest location.

Supernova and host galaxy sketched on a log sheet and marveled over for quite some time, I return to Leo. I know I haven't exhausted all his wonders, not by a long shot. Not all my targets are easy from the city, and I have failures as well as successes, but instead of bemoaning my horrible sky, I simply resolve to revisit the "not-seens" again on a slightly better night.

As the evening grows older, I see the lights in the house begin to wink off as my wife prepares our home for deep night. The door opens and she walks lightly into the backyard, wanting to spend a little time with me as the day enters its dark, quiet reaches. I turn the scope back to the supernova, and we admire it together, wondering softly aloud. Then we step back from the telescope and just contemplate the eternal stars together. Neither of us notices the ugly light pollution, really, we simply appreciate the beauty we're given in silence until we're startled by the "WHOOO!" of the neighborhood owl who's winged in, wondering what we're doing—or maybe just looking for a stray mouse.

Beyond the many sky marvels I find on every city night, there are the practical pleasures of using an urban site. When it grows late and I know it's time to call it a night, it takes all of 10 minutes to carry scope and accessories back into the house and be drinking a whiskey, ruminating on the sights I've seen over the last several hours. The ease with which I can set up and teardown means I'm not only anxious, but *eager* to observe on every clear night. While the once-a-month dark-site-only observers are complaining about what a terrible spring it's been weather-wise, I'm remembering the *many* nights I've spent with the deep marvels of Virgo and Leo.

Finding an Urban Observing Site

Before you can start taking advantage of urban observing, you've got to have an observing site. If you live in a detached home with even a small front or back yard/garden, your problems are over. Like me, you just trot the scope out the door and start having fun. If you live in an apartment or townhouse, however, the solution is not quite so simple. One alternative for the apartment resident is the roof. Often the roof of an apartment building is accessible by an elevator or stairs if you're lucky, or a ladder and hatch arrangement if you're not so lucky. If you're faced with the latter, the best you can hope to do equipment-wise is a small refractor or binoculars—you're not going to lug a 16-inch Newtonian monster of a telescope up a ladder. Even if all you can use on the roof is your Short Tube 80-mm refractor, though, the experience is usually going to be a nice one.

"Up on the rooftop" you've likely got fairly unobstructed horizons, and, assuming the roof area is not lit by the all too common mercury-vapor security light, you may find a little relief from light-pollution up there as well. At least you'll be able to avoid a lot of the ambient light at street-level. Naturally, before you start using the roof it would be wise to inquire as to the feasibility of doing so with your building superintendent. You don't want to suffer the ignominious fate of being locked out while up on the roof some night, and you certainly don't want to do something that would endanger your lease.

What if the roof is inaccessible or otherwise impractical for use as an observing platform? In some areas of the world, especially the older parts of larger European cities, flat-roofed apartment houses are uncommon. Or what if you live in a townhouse or other attached single family dwelling without a usable roof area *or* a yard? If you have a balcony, that will provide you with a usable, albeit cramped and limited (in the amount of sky you can see), observing platform. Actually, you'll be surprised at how much you'll see, even in the limited expanse of sky offered by a balcony if you're patient and observe at various hours of the evening as the seasons progress. But you may want to search for an alternate site, one you can use on the occasions when you need to see a part of the sky invisible from your balcony roost.

If you have neither usable roof nor garden and no balcony either, your best bet may be to discuss your problem with the local astronomy club. Chances are, they know of safe and convenient areas where you can observe in town. Even the largest and most light-polluted metropolises have active astronomy clubs whose members observe from within the city limits at least part of the time.

What are possible observing locations other than home? A school or science museum with a flat roof or secure parking lot or other open area is a good alternative. Frequently, these institutions will be willing to provide you with observing space if you'll agree to help them with public outreach astronomy activities once in a while, especially if you approach them with your astronomy club friends as an organized group. The problem here is that most of the open areas possessed by city schools, museums, and similar organizations will be heavily lighted with the brightest sodium or mercury vapor lights money can buy. Sometimes these can be turned off, often not. Even in the case of constantly burning security lights, though, you will probably find at least one shadowed corner where you can observe profitably.

Parks and other public areas are another possibility, but a couple of difficulties exist with these. Most limiting will be the city's rules concerning your use of these locations. In my town, for example, there's a beautiful and safe municipal park that would provide a good observing venue. Unfortunately, despite few demonstrable problems with anybody in the park over the years, the City Fathers in their wisdom have seen fit to close it at sundown, pretty much eliminating it as a "legal" observing site. Even if you are allowed to use a park after dusk, there is a very important concern when considering parks and other wooded urban areas as observing sites: your personal safety.

Security and the City Lights Astronomer

My astronomer friends who live in the country are always surprised that I'm not "afraid" to observe in the city. I find this a little funny, as the only times I've felt overly nervous while observing have been when I've been alone at a location far out in the country. I know what to expect in the city, and, whether at home or at one of my other in-town locations, I've never felt anything but safe. There is no doubt that safety *is* an issue for urban astronomers, though; particularly those who choose to observe from public-accessible sites where the very things that make the location attractive—fewer streetlights and the presence of wooded areas to block stray illumination—may cause genuine safety concerns. If you choose to use a public area, the first step to safety is in *knowing your observing site*. Is there a genuine crime problem? Are there gangs or homeless persons in the park after dark? If, after checking the park or other location personally (after dark) and perhaps talking to area residents, you turn up any "yes" answers, I would discourage you from attempting to use said site—alone, anyway.

Even if you judge an urban park or other public site "safe," you should still keep security in mind while planning and conducting observing sessions. There are some things you can do to help ensure your safety when observing away from home in the city (or, really, anywhere else). The first is to use the buddy system. If you have an active, enthusiastic friend, taking her or him along with you on your observing expeditions can go a long way toward ensuring your safety. A couple of friends is much better. I think it is *always* wise to observe with a companion when away from home, no matter how supposedly secure your site. Criminals and crazies will almost always be less than anxious to take on a group, but may see a lone person as "prey." Also very important: always let someone know where you will be and when to expect you back.

"What can I take with me when observing to help keep me safe?" When my fellow American amateur astronomers ask me this question, it's usually a polite and roundabout way of asking whether I think they should carry firearms when they observe. In my younger days, I would sometimes take a handgun with me—when observing alone far out in the country, never in the city—but I never, ever, had recourse to use my "piece." *Not even close.* In my opinion, a gun, a dark site, and a nervous astronomer, especially one not overly familiar with firearms and firearm safety, can be a recipe for disaster. Very easily. Other reservations aside, a gun is not a solution for a very good, practical reason: *if you are so nervous about your safety at an observing location that you feel the need to pack a firearm, you will most certainly not be able to do any fruitful observing. You'll be too nervous to concentrate, and will be jumping at every sound.* Forget guns. For most of the world's amateurs, especially those in the UK and Europe, a firearm isn't an option anyway. But there is an item you *can* take along to help ensure your safety, a cell phone. The cell phone, in my opinion, is a must for anyone observing alone anywhere, and is much more useful than a pistol. The gun won't be much help in the event of a dead car battery!

For the urban observer (or the suburban or country observer) the real way to safety is, again, the choice of a safe, comfortable site. If you feel secure, observing will be much more fun and you'll get a lot more accomplished. Of course, in the dark hours of the night it's easy to get spooked at *any* site. I recall one late evening in my familiar, safe, fenced backyard when I started hearing noises. Snapping twigs every once in a while. Eerie sounds of rustling leaves. Just as I was ready to run for the back-door, the psycho killer-UFO alien-werewolf turned out to be a friendly opossum, a common member of the urban fauna here, stopping by to say "hello" and see if I'd hand out some food.

The urban astronomer faces another security concern that's not related to the bad guys. It's the *good* guys. An amateur astronomer, either alone or with companions, is of immediate interest to passing police officers. This is understandable. You're out there alone in the shadows with a thing that, to the non-astronomy-literate law enforcement person, looks suspiciously like a *weapon* of some kind. Maybe a rocket launcher or a cannon. This would have seemed ludicrous a few years ago, but now, especially in the suddenly very security-conscious U.S.A., it is a very real scenario (understandably). Not just in the States, either. From talking to my astronomer friends all over the world, I conclude that it's not at all unusual anywhere for Joe or Jane Amateur Astronomer to be quietly admiring the heavens when the entire universe is suddenly illuminated with flashing lights and a stern voice intones, "Don't move!"

The secret to surviving these encounters in one piece and with your sanity and freedom intact is to do *exactly* what the officer instructs you to do. Assuming you're not some place you are not allowed to be, the policeman will usually end up being apologetic and will happily accept a view through the telescope (maybe, secretly, for final assurance that it's not *really* an ICBM launch tube). What can you do to avoid these encounters? If you observe from your home, let the neighbors know what you're doing: you're looking at the stars, not their bedroom windows, and it's you out in the yard with that funny tube, not a terrorist nutcase. If you're in a public area, make sure it's a location where nighttime access is allowed.

Let me emphasize this again, if confronted by the police, keep your cool and follow their instructions *to the letter.* Honestly, they have the right and the reasons to be curious and concerned about anything unusual they encounter on the urban landscape. Don't be scared off from your legal observing site, though. If the police seem to be hinting that you need to "move along," politely remind the officers that you're lawfully enjoying the park (or other location), just like the couple necking on the bench down the way.

Evaluating a City Observing Site

Now that you have a safe and convenient urban site to use for your observing runs, what can you *expect* from it? How bad is "bad" when it comes to observing the deep sky? Sky glow is a given. No matter where you go in the city, the sky is going to be bright due to the presence of thousands of unshielded or poorly shielded lights. You can't do anything about that. Your main concern is the *other* part of the city light pollution equation, the part you *can* do something about, ambient light. Ambient light is the stray light from nearby fixtures that's shining directly into your eyes. In some ways it is even more harmful than sky glow, since you could see a lot more, even in your city's compromised sky, if your eyes could gain some measure of dark adaptation. With a brilliant security light shining straight into your face, your pupils will remain as constricted as they can be, and even a bright open cluster will be hard to see. If you must choose a site that's badly affected by ambient light, there are ways to block it from your view, as we'll see in Chapter 3, "Accessories for Urban Observers." It's best, however, to seek a site that's shielded in some way from direct light if at all possible. A building, a tree, or a simple light shield (also in Chapter 3) can improve your ambient light situation a lot.

You'll also want to know the *limiting magnitude* of your city and your site. "Limiting magnitude" simply means, "How dim are the stars I can see with my naked eye?" In dark country you may be able to "see" down to magnitude 6 or even 7, which will mean the sky is festooned with countless of stars. From a heavily light-polluted city, you'll probably be limited to magnitude 4 or 3 stars. It's rare for things to be much worse than that, as, at magnitude 2, even a bright star like Polaris is barely visible. Possible in the largest and most light-polluted cities, but not likely. If you can't see a second magnitude star from your site, the likely cause is *ambient* light keeping your pupils "stopped down." A location that will allow you to see magnitude 4 stars away from the bright horizon will be a very good site and will provide countless hours of star-gazing enjoyment. Even a magnitude 3 site is quite usable, especially on evenings when the sky is dry and clear or if you restrict your observations to areas approaching the zenith.

How do you determine your limiting magnitude? It's easy. Find a constellation that's well away from the horizon and note the dimmest stars you can see. Away from the horizon because light pollution is always at its worst near the horizon. The poor atmospheric seeing, dust, and thick air there mean you won't want to observe objects below about 30° in altitude, anyway. A traditional tool for determining limiting magnitude is the Little Dipper, Ursa Minor (Figure 1.1). It provides a good spread of star brightnesses from magnitude 2 on down to magnitude 5 in a relatively small area of the sky. For best results, wait until all parts of the dipper are well away from the horizon before you use it. Also try to wait for a night that's pretty average as far as transparency and humidity go (high humidity skies scatter light and make existing light-pollution worse) so you get a good idea what to expect most of the time. Once

Ursa Minor

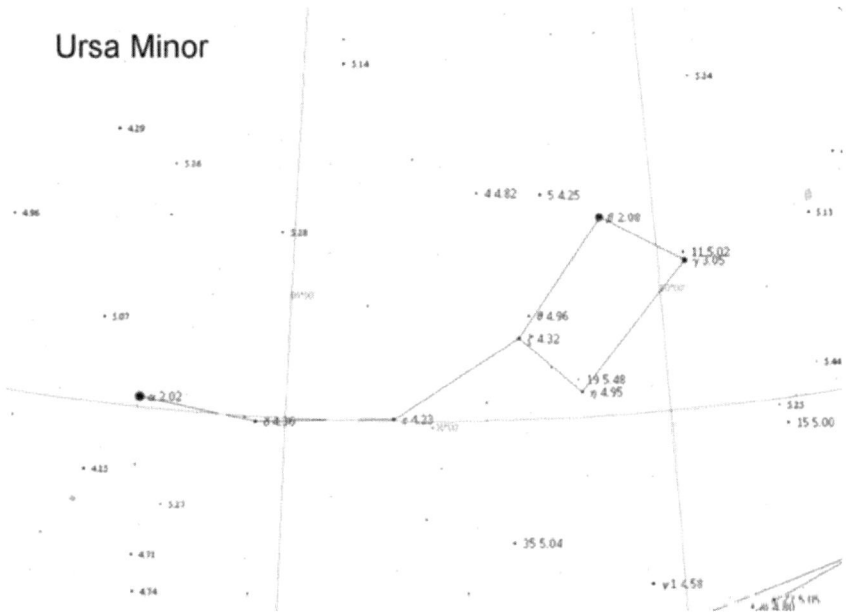

Figure 1.1. Use Ursa Minor to determine your limiting magnitude.

you know the condition of your sky, you'll be able to choose appropriate objects for an evening's observing program and will know, to some extent at least, how hard a "DSO" (deep sky object) will be to track down and observe.

Once you know *where* to observe, what do you observe *with*? Telescope choice is important for any amateur, but choosing the optimum instrument is critical to your enjoyment of the urban sky. The following chapter will help you select a telescope to serve as your urban starship.

CHAPTER TWO

Telescopes for Urban Observers

What's the best telescope for the city-bound deep sky astronomer? If you already own an instrument, *that's it*. Almost any telescope, perhaps with a few simple modifications, can work well as your urban starship. Not all telescopes are created equal, however, and if you're thinking of expanding your telescope arsenal or buying your first instrument, you now have the chance to acquire one with your personal observing environment in mind.

Telescope Types

Before you can select a telescope in an educated manner, you need to become acquainted with a few of its characteristics. There are three basic designs: the refractor, which uses a big lens (usually referred to as the "objective") to collect light; the reflector, which uses a large concave mirror (the "primary" mirror) for the same purpose; and the catadioptric telescope, which uses a combination of lenses and mirrors to grab starlight (Figure 2.1 shows the most common designs). Two simple specifications will tell you a lot about a telescope of any design.

The first specification is *aperture*, the diameter (expressed in millimeters for small scopes and sometimes inches for larger ones). This indicates how much light the telescope can collect. *Light* is what you want, whether you observe from the city or the country. Any scope can be magnified to any extent. Plenty of light is what's needed, not the higher magnification. The department stores are filled with "600 power" 60-mm aperture scopes in alluring boxes festooned with Hubble Space Telescope images. Some of these scopes are actually fairly good optically, but, unfortunately, they are completely useless at high magnifications claimed for them. High power with a small

Lens Type (Refracting) Telescope

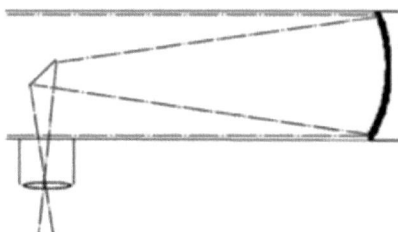

Mirror Type (Reflecting) Telescope

Lens - Mirror Type (Catadioptric) Telescope

Figure 2.1. Major telescope designs.

telescope makes everything dim to the point of invisibility. Images in the eyepiece must be bright enough for high power to be useful. A deadly dim globular cluster at 300× will show the observer *less* than what he or she could see at 100×. Light gathering power depends on the area of the lens or the mirror, so an objective lens or primary mirror with twice the diameter of a smaller one will collect four times as much light.

The second important specification is the *focal length* of the scope, the distance from the lens or mirror where the image comes to a focus. It is commonly expressed in millimeters, even if the mirror size is given in inches (don't ask me why). Longer focal length telescopes deliver higher magnification for a given eyepiece. A scope with a focal length of 750-mm, for example, will provide 30× with a 25-mm focal length eyepiece, whereas a telescope with a focal length of 1,200-mm will provide a magnification of 48× with the same 25-mm eyepiece (magnification can be calculated by dividing the focal length of eyepiece by the focal length of telescope).

Focal ratio is very similar to focal length. It is the mirror (or lens) aperture divided by the focal length of telescope. A telescope with a 150-mm mirror and a focal length of 1,200-mm has a focal ratio of *f*/8. Similarly, a 150-mm aperture telescope with a focal length of 750-mm yields, a focal ratio of *f*/5. Smaller focal ratios for a given size of objective mean shorter focal lengths, lower magnifications, and wider fields.

Larger focal ratios denote larger magnifications and narrower fields. These focal ratio numbers will soon become second nature to you when it comes to evaluating telescopes. If you see "$f/4$," you'll think "low magnification and wide field." A focal ratio of "$f/10$" will mean "high magnification, narrow field." With these few simple scope characteristics in mind, you're almost ready to start considering "which scope" in detail. Before looking at urban telescope candidates, however, I want to put to rest an old and silly myth.

The Urban Aperture Myth

You've heard it before, from local amateurs, on Internet astronomy discussion groups, and even from prominent astronomy authors who should know better: "If you live in the city or heavily light-polluted suburbs, don't buy a large aperture scope. A big mirror or lens will collect more light, but this will include more sky glow, more light pollution. The sky background in a big scope's field of view will be so bright that you'll see more with a smaller instrument. Get a 4-inch refractor, not an 8-inch Schmidt-Cassegrain or 12-inch Dobsonian reflector." Sounds reasonable and sensible. Big scopes gather more light, both from distant deep sky objects (DSOs) and from the background sky glow in your light-polluted skies. Choose a nice, small scope instead.

The problem with this theory is that it is a nonsense. Even though a big scope does collect more light from the bright sky background, *its deep sky images always look brighter and more detailed.* In order to prove or disprove this urban-aperture-limitation theory, I set up a small aperture 4.25-inch Newtonian reflector next to my largest scope, the 12.5-inch Dobsonian. I then pointed both at M13, the marvelous globular star cluster in Hercules. Assuming the urban-aperture theory was correct, the views in both instruments should have been similar. The 12.5-inch Dobsonian would yield an image so washed out by bright sky background glow that no more details would be visible in the cluster than in the 4.25-inch Newtonian reflector.

When the smaller scope was pointing at M13, I inserted an eyepiece that yielded 48× and took a look. It looked nice! The great cluster was easily visible, bright, and seemed as if it might *want* to resolve into myriad stars with higher magnification, but I wasn't able to see any individual stars at 48×, not even around the cluster's edges. What would I see in the 12.5-inch scope? I moved the eyepiece to the larger scope, where it gave the roughly comparable magnification of 65×. WOW! M13 didn't just look *nice*—despite my less than dark skies, it was a *marvel*. Many, many tiny cluster stars were visible, and, with the globular riding high in the sky, I seemed to resolve it across its very core with averted vision. It wasn't just a round glow; it was a big ball of stars. What a difference!

Maybe the comparison wasn't exactly fair? The 12.5-inch scope's slightly higher magnification could have given it an overwhelming advantage. I searched around in my eyepiece box and came up with a longer focal length ocular that gave me a magnification of 45× in the 12.5-inch instrument. Nope! The view in the 12.5-inch scope was still better, much better, than the attractive but unresolved view in the 4.25-inch. Frankly, in the 4.25-inch scope, M13 looked like a fairly unimpressive smudge. A bright smudge, but a smudge nevertheless. Also, to my eye, the field background really didn't look much brighter in the 12.5-inch scope than it did in the 4.25-inch. The background was bright in both instruments, but, to me, not noticeably moreso in the larger instrument.

Maybe the aperture gap was just too great. Perhaps an 8-inch would be a more worthy opponent for the big 12.5-inch than the little 4.25-inch in the city? I set up an 8-inch $f/7$ Newtonian reflector that I had on hand and took a look at Hercules. The cluster was better than it was in the 4.25-inch scope, but the view was not nearly as good as in the big 12.5-inch Dobsonian. M13's appearance in the 8-inch was considerably better than it was in the 4.25-inch scope, though—some stars were easily visible. The conclusion was unavoidable. In the city, as in the country, *aperture wins*. The larger your lens or mirror, the better the view.

When people ask me about the urban-aperture myth these days, I reply, "If you observe in light-polluted areas, always choose the largest aperture telescope you can afford and transport. In the city, aperture always wins." In fact, I've come to believe that aperture is more important in the city than it is in the country or suburbs. From a dark site, a surprisingly small scope will show a lot of deep sky objects in detail. At a pitch-black desert location, my little 4.25-inch reflector would undoubtedly have done better on M13 than it did in the city. No, it still wouldn't have kept up with the 12.5-inch scope, but the cluster would've looked better; some stars would have been resolvable.

If the sky is bright, you need all the aperture horsepower you can muster. Don't let anybody convince you otherwise with tales about the "bright sky background." If the field in your larger aperture telescope looks annoyingly bright, increase the magnification—that will darken it. But at *any* magnification, deep sky objects will show more detail with large aperture than with a small scope.

How Much Aperture?

The foregoing would seem to eliminate small telescopes for city use. That's not strictly true. Large aperture is always best, all things being equal, but all things are *not* usually equal. My big Dobsonian-mounted Newtonian works for me, living in a single-family home with a backyard, but if you live on the 12th story of an apartment building, you'll find hauling a big scope up and down a bit troublesome in an elevator and completely impossible if you have to negotiate stairs at any point—to access your building's roof, for example. If you're in a situation where you have to transport the scope to observe, pick a telescope with only as much aperture as you can *handle*. Try not to go too small, though.

If you can gain a sizable performance increase by going to a larger telescope that's only slightly more difficult to move, by all means do so. For example, given the choice between a 4-inch medium focal length refractor and a 6-inch Newtonian reflector, I'd always choose the 6-inch Newtonian reflector. The 6-inch reflector is slightly more difficult to move around than the 4-inch refractor, but only slightly, and it's worth the extra trouble for those additional 2 inches of aperture. Since it's an area that counts, the jump from 4 inches to 6 inches makes a big difference in what you can see. You'll hear a lot about the superiority of refractors with respect to contrast and image sharpness. Some of what you hear is true, but for the urban deep sky observer, again, the prime requirement is *light*. A 6-inch reflector will deliver more precious light than the 4-inch refractor.

How Much Focal Length?

Given the choice between a short focal length, small focal ratio scope, say, a 6-inch $f/4$ (focal length 600-mm) and a 6-inch $f/8$ (focal length 1,200-mm) for use in the city, I'd pick the $f/8$. Why? Larger telescopes are not handicapped by bright city skies any more than smaller telescopes, but all telescopes are naturally troubled by the relatively bright background of a low-power field delivered by scopes in light-polluted areas. In the country, nothing is nicer than touring the heavens with a low-magnification eyepiece. The sky background is velvety black and objects stand out in stark relief. In the city, the sky in your eyepiece is light gray rather than black. There's less contrast between the sky background and deep sky objects, and some dimmer DSOs may disappear altogether. Luckily, we can combat this bright field effect. Higher *magnification* increases the *contrast* between an object and the sky background—which is not to mention that you can duplicate country conditions by using high power, but it does help.

Why choose a larger focal ratio and longer focal length telescope for the urban use? A larger focal ratio scope produces *higher magnifications* for any given eyepiece. It's easier to reach a usable magnification for the city with common eyepiece focal lengths with a larger focal ratio scope. You don't have to resort to short focal length eyepieces—which are often uncomfortable to use—to achieve the higher powers you need, as you may have to do with a small focal ratio, wide field telescope.

Another benefit of large focal ratio instruments is their optical quality. Large focal ratio optics are always easier to make than smaller focal ratio optics, so, for a given price, a larger focal ratio scope may be considerably better optically than a smaller focal ratio scope.

A small focal ratio telescope may be desirable for the urban observer for easy portability, however. Smaller focal ratio, shorter focal length refractors and reflectors have shorter, lighter tubes than large focal ratio, long focal length instruments, making them easier to transport and store, which may be critically important for the apartment dweller.

Which One?

A quick browse through the amateur astronomy magazines will reveal a multitude of advertisements for a bewildering array of telescopes. The prospective 21st century telescope buyer is lucky that there is so much to choose from, but the endless color ads and manufacturers' enthusiastic claims and counterclaims make the task of picking a telescope confusing and maybe just a little bit frightening. The following section is designed to make this process less scary. In the next few pages, I'll look at the major telescope types with an eye toward their suitability for use in the city. I'll also consider some specific models. Since there are more commercial telescopes available today than

I could possibly provide educated hands-on reports for, the fact that a certain brand or model is not represented here does not necessarily mean that it is a bad scope or a bad scope for the city. It may just mean that I haven't gotten around to trying it. But the listed telescopes are my favorites and ones that I've had the chance to use in the city—often extensively.

The Refractor

Prior to about 20 years ago, the refractor, the time-honored "big lens" telescope, was dead when it came to amateur astronomy. Refractors, once much-loved by amateurs, had, with their big price tags, colorful images (as in chromatic aberration) and long, unwieldy tubes been left in the dust during the 1970s and 1980s by Schmidt Cassegrains and big Dobsonian (Newtonian) reflectors. But the refractor has staged a remarkable comeback, and it is now once again a popular and logical choice for any amateur and certainly for the city observer.

What brought the refractor back? Three things. First, the premium color-free apochromatic objective lens designs pioneered by Astro-Physics and TeleVue in the United States and Takahashi in Japan. Refractors suffer from chromatic aberration—a problem that's plagued these scopes since Galileo's day. An achromatic refractor, i.e., a telescope with a two-element objective made of crown and flint glass, the most popular design of refractor objective lens for the last couple of centuries, cannot bring all colors of light to focus at the same point. The practical effect of this is that bright objects like the Moon, Jupiter, Venus, and brighter stars are ringed with purple halos. This "excess" color is not only distracting—for some observers, very distracting, as some people seem more bothered by chromatic aberration than others—it tends to obscure detail and soften the image.

The apochromatic ("without color") refractor solves the chromatic aberration problem. Sophisticated lens designs and innovative materials—the use of fluorite "glass" is common—make the "color purple" a thing of the past. The color-free nature of these telescopes allows them to show the refractor's strengths to best advantage: good contrast due to the lack of a central obstruction (from a secondary mirror), a maintenance-free sealed tube, and little need to allow the telescope to adjust to outdoor temperatures. Couple this with the mechanical perfection the APO makers lavish on their *beautiful* creations, and you have very capable telescopes with heirloom quality. Naturally, this comes at a high price. Apochromatic telescopes are the most expensive telescopes per inch of aperture, and the prices really skyrocket once you get above the fairly modest aperture of 5 inches.

The second reason refractors returned was an emphasis on smaller focal ratios. In their earlier incarnations as amateur telescopes, most "featured" focal ratios as large as $f/16$. Few were seen "faster" (i.e., with smaller focal ratios) than $f/12$. This was done in an effort to reduce chromatic aberration, as at large focal ratios with their resulting long focal lengths, color is reduced for any achromatic objective. In theory this is a good idea, but at these very large focal ratios, once you got above a few inches of aperture, fields were very small and magnifications very high for any given eyepiece. Photography of deep space objects was difficult due to the very long exposures such "slow" optical systems required. In addition, the tubes were long and awkward (almost

6 feet long for a 4-inch telescope), and discouraged the user from moving the scope often, whether to dodge lights or to travel to dark sites. All in all, these telescopes were for an older version of amateur astronomy, one that focused on Moon, planets, and a few bright DSOs.

Today's refractors, and especially the apochromats, offer up delicious wide fields and fast focal ratios that make them attractive to deep sky fans. With focal ratios from $f/5$ to $f/6$ typically, a stunning view of the Pleiades or the whole sword of Orion is possible in the smaller apertures. This smaller focal ratio popularity has also affected the achromats, which now hover around $f/8$–$f/10$, with $f/5$ s also popular. It is almost unheard of to find any type of refractor with a larger focal ratio than $f/10$ these days. This works very well for the apochromats; they can deliver low powers and wide fields but, due to the superb quality of their optics and the lack of color problems, they still allow high magnification viewing. Achromats with shorter focal ratios are less successful. Even at $f/10$, color is quite noticeable, and, due to chromatic aberration, higher magnifications cause rapid breakdown of image sharpness.

The final key to the resurgence of the lens-scope? The influx into the West of very inexpensive but relatively well-made Chinese (Mainland and Taiwan) refracting telescopes. At this time, most of the Chinese refractors are traditional achromats, but they are relatively well made and perform well considering their low prices. These low prices have allowed many of us to enjoy an experience that, for those of us who entered astronomy in the 1960s or before, always seemed impossibly expensive: owning a "big" 6-inch refractor.

Are refractors a good choice for the urban astronomer? Yes and no. They are a very good choice if portability is a major concern. An $f/5$ or $f/6$ refractor in the 5-inch or smaller aperture range is easily manageable, even if the owner must reach her viewing site via multiple flights of stairs. The problem comes when it's time to increase magnification. Urban observers often use higher powers than country-based astronomers. As explained earlier, this darkens the field of view and allows DSOs to pop out of the normally gray sky background. If the $f/5$ or $f/6$ telescope in question is an apochromat, no problem. Just run the power up as much as you want. But be aware that you may have to use very short focal length eyepieces to reach this desired power.

If the refractor in question is a small focal ratio achromat, however, you may find it all but unusable in the city. A 4-inch $f/5$ Chinese achromat, for example, will often "top out" at around $100\times$. At higher magnifications, images become impossibly blurry due to a variety of optical imperfections, and that is a big problem in the city, as $100\times$ is often not nearly enough power.

Does that mean urban astronomers have to pay for an apochromat if they want a refractor? Maybe not. My tests have shown that at least some bargain Chinese achromats can be pushed to higher powers if you are content to stay away from the Moon, planets, and bright stars. Avoiding the bright, color-plagued objects helps these scopes keep it together magnification wise a little longer. But even on dimmer DSOs there's no avoiding the fact that the short achromats deliver their sharpest images at low powers.

If there's one thing that argues against the refractor, achromat or apochromat, as an urban astronomy tool, though, it's aperture. 6 inches, 150-mm, of aperture is about where the city astronomer needs to *start*. That's big enough to start pulling some interesting objects out of the light pollution. 8 inches, 200-mm, of aperture is even better. Unfortunately, even a lightly built Chinese 6-inch refractor is B-I-G. It's not something you'll want to waltz from one side of the yard to the other to avoid porch lights, much less carry down four flights of stairs. Premium 6-inch refractors are even worse. They and their mounts are large and very expensive. An 8-inch apochromat is not only huge (as a portable scope), it's hugely *expensive* for most amateurs. In contrast, an 8-inch Newtonian reflector is relatively light, inexpensive, and especially on the deep sky, can deliver *most* of the image quality of an APO costing 10 or 20 times as much.

Refractors have a lot of charm, and I wouldn't fault you for choosing any of the scopes listed below, but for those of us condemned to do most of our observing under the glow of sodium streetlights, there are arguably better choices.

Synta

While most of the telescopes in our survey are sold under a single name, the Synta refractors are confusingly offered wearing many badges. These popular telescopes, both "short-tube refractors" and longer focal length scopes, all made by the same Taiwanese firm, are widely available in the United States and Europe under numerous brand names, with "Skywatcher" and "Konus" being plentiful in Europe and the U.K. and "Orion" and "Celestron" being the name plates on most U.S. models. They are all very much the same with typically the only difference from one brand to the next being the color of paint on the tube.

Synta's Short-Tube Refractors

Synta produces a full line of short focal length achromatic refractors, all with focal ratios of $f/5$. In addition to the original short tube, the Short Tube 80, an 80-mm $f/5$ refractor, which was the first Synta to become popular with U.S. amateurs, the company also offers a 102-mm $f/5$, a 120-mm $f/5$, and a (seldom seen in America) 150-mm $f/5$. How good are these scopes? Pretty good, considering their low prices. In some ways the best of the lot is still the 80-mm $f/5$ (Plate 2). Its smaller objective with its smaller amount of chromatic aberration and resulting ability to take higher magnifications than its bigger brothers makes it more versatile. It's a little handicapped for city use because of its small aperture and small focal ratio, though. The larger models display more excess color and are less able to handle higher magnifications, but, as mentioned earlier, staying on the deep sky enables them to do higher powers more gracefully than if you attempted the Moon and planets, subjects for which they are not well suited.

Synta's Medium Focal Length Refractors

Synta is famous for its short tubes, but its medium focal length telescopes are nearly as popular, with the 102-mm $f/10$ refractor a close second in popularity to the 80-mm $f/5$. All these refractors have focal ratios close to $f/10$ and all come equipped with workable German equatorial mounts, the medium-sized EQ3 for the 102-mm and the EQ4 for the larger models. In addition to the 102-mm, Synta offers a 120-mm and a 150-mm refractors.

All of Synta's larger focal ratio refractors with their resulting longer focal lengths are reasonable choices for urban use. One thing the prospective buyer should remember, however, is that the focal ratio of $f/10$ is still small when it comes to reducing color in achromats. Focal ratios would have to be half again as large for color to begin to disappear, especially with the 120-mm and 150-mm telescopes. These refractors can do a good job on the deep sky, however, with only the brightest stars showing much disturbing color.

Meade

Meade's Achromatic Refractors

The U.S. company, Meade Instruments Corporation, is most well known for its Schmidt Cassegrain telescopes (SCTs), but it also offers some refractors of interest to the urban astronomer. Once you get beyond the small, cheap department store scopes Meade imports and sells, you are left with three interesting achromats. The ETX 80 is a short focal length $f/5$ refractor on a computerized fork mount. Equipped with its Autostar controller, this little wonder will automatically find over a thousand objects. In another class altogether are the AR5 and AR6 refractors (127-mm and 152-mm apertures, respectively). These telescopes are part of the company's LXD-75 series, and are mounted on German equatorial mounts that, with the included Autostar computer, will automatically point them at tens of thousands of objects. Naturally, only a small subset of these will be visible from our light-polluted haunts.

The ETX 80 is a cute little scope. I own and use the very similar ETX 60 (now discontinued), and have had a lot of fun with it. Unfortunately, its short focal length means that, like the Synta short tubes, it's best suited for dark skies and wide-angle DSO viewing. In my experience, it can beat the Synta 80-mm $f/5$ in the city, since the go-to feature means that it's easier to locate objects in bright skies. Image quality is similar to the 80-mm in brightness, but the 80-mm $f/5$ s usually do better on the planets.

The AR5 and AR6 are interesting for a couple of reasons. First, their prices are surprisingly modest. They also possess apertures that move into the range suitable for deep sky viewing in the city. Most observers have given good reports on the optics that, while imported, seem slightly better on average than those in the comparable apertures Synta telescopes. The Chinese-made LXD-75 GEM mounts seem well-thought out if a little rough around the edges, and the computer features really work if the telescope is carefully aligned.

Meade's ED Refractors

Meade also offers a line of "almost" apochromats, the ED series. They are more expensive than the achromats, but much less pricey than true apochromats from premium class telescope makers. These telescopes, available in apertures from 4 to 7 inches, as tube assemblies only or on computerized mountings, have been reasonably popular with amateurs. With the exception of the 4-inch, they are a little heavy for the average urban observer, though, and, mechanically they have had a few problems.

The focusers on Meade's inexpensive AR scopes are considerably better, for example, than those on the more expensive ED models. Meade has done little advertising and promotion of these instruments in recent years, and, while they are still offered, I'm not sure how much longer they will be around. I expect them to be replaced by comparable aperture Chinese ED semi-apochromats "any day now."

TeleVue's Telescopes

TeleVue is justly famous among amateur astronomers for its premium wide-field eyepieces (see Chapter 3), but it also produces refractors second to none. The TV line starts out with the 70-mm Ranger and Pronto. These both telescopes use the same 70-mm *achromatic* objective (albeit with ED glass elements), but differ on mechanical features, with the Pronto, which features a 2-inch focuser, being heavier and more expensive. Both the Ranger and Pronto have been discontinued recently, but are still available from many dealers.

When it comes to genuine apochromatic refractors, TV offers the $f/6$ TV60 (60-mm), the $f/6.3$ TV76 (76-mm), the $f/7$ TV85 (85-mm), the $f/5.4$ NP101 (101-mm), and the $f/8.6$ TV102 (102-mm). All the APOs feature 2-inch focusers and premium fittings. TeleVue sells a line of alt-azimuth mounts in various sizes, but the telescopes are normally sold without mountings, allowing the user to choose one that best suits her/his needs.

The TeleVue APO refractors are excellent in every way. The images produced by those I've used have been as good or better than those of any apochromat I've tried. There are a couple of drawbacks to these fine scopes, however. First, there's the price. Like all top-of-the line refractors, they are expensive, with prices ranging from 700 US$ for the Ranger to 3,600 US$ for the NP101. Remember, these figures do not include the extra expense for mountings. The lowest priced TeleVues, the Ranger and Pronto are achromats. They may be somewhat better than the average Synta Short Tube when it comes to chromatic aberration, but will not be color free.

You do have to expect to pay for perfection, of course, and that's what the TVs deliver. But there's a second and more serious quibble. A 4-inch is a 4-inch is a 4-inch. No matter how finely crafted a telescope is, aperture is *still* the key, no matter what the design. A 4-inch is *usable* in the city, but more aperture is better. The 101 and 102 scopes are the *minimum* for serious urban work. If you can pay the fare, though, go for it. A TV101 or 102 on a comfortable alt-azimuth mount and equipped with a digital setting circle (DSC) computer (also available from TeleVue) is a portable setup that will last a lifetime and one that may surprise you with its performance, even under city lights.

Astro-Physics

If any name is associated with the rebirth of the refractor, it's "Roland Christen." Mr. Christen, the founder of the U.S. firm Astro-Physics Incorporated, "AP," has been turning out world-class refractors for over 20 years. His telescopes are considered by many amateur astronomers to be the best in the world. Currently offered are the $f/7$ Stowaway (92-mm), the $f/6$ Traveler (105-mm), the $f/6$ Starfire (130-mm), and the $f/7$ Starfire (155-mm). The Starfires are all literally *color free*, and the color error in the smaller scopes is so small as to be of technical interest only. In addition to the telescopes, Astro-Physics also sells their own line of computer-equipped go-to German equatorial mounts.

Can anything bad be said about Astro-Physics telescopes? Not really. For what they are, they are exquisite. Naturally, as with the TeleVues, this comes at a price. While very reasonable when compared to other premium brands, the APs don't come cheap—the 92-mm costs 2,880 US$ and the top-of-the-line 155-mm quipped with all the options commands as much as 8,800 US$. On suitable mounts, the 5- and 6-inch refractors, like all examples of their breed, are less than portable, and, in most cases, not something the apartment dweller should consider. Aperture is a problem, too. A 6-inch aperture is good, but not great when you're faced with bright skies. Finally, you can't just go out and buy an AP. To get one, you'll wait three years or *longer*. Despite their prices the scopes are very popular and are produced in very limited numbers. "AP" is currently in the process of switching from the above telescope models to some new designs, and that may possibly result in even longer waiting list for these super-premium telescopes.

Others

For the "price is no object" crowd there's Takahashi. These Japanese APOs are very close in image quality to the TVs and APs. The trade-off is that they are considerably more expensive. They are available immediately—no waiting list—however. An up and coming company in the U.S. is "TMB Optical." This concern, named after the initials of owner/designer Tom Back, is producing telescopes that are very well regarded by amateurs, are comparable in quality to the rest of the premium pack, and, at this time, are available a lot sooner than the APs. For the achromatic refractor enthusiast, D&G Optical (U.S.) makes achromats in a variety of focal lengths (including long ones) that are in the premium class compared to the humble Syntas.

Newtonian Reflectors

If you ask an advanced amateur astronomer which telescope is "best" as a first, serious purchase, 9 out of 10 times the reply will be "Newtonian reflector." These simple, inexpensive telescopes do have a lot to offer observers—in or out of the city. Their main strength is certainly their dollar/aperture ratio. When it comes to aperture for your telescope money, nothing beats a Newtonian. There are some expensive, premium examples, but it is quite possible to get a working 16-inch Newtonian for just over 1,000 U.S.$. This design is also quite versatile; a properly made Newtonian with a well-made primary mirror is capable of handling a wide range of magnifications and delivering outstanding images of planets and DSOs.

Potential problems with Newtonians for the urban astronomer? The lower priced models are Dobsonians, "dobs" (named after their popularizer, John Dobson), which are Newtonians mounted on simple alt-azimuth up–down/left–right mounts usually made of wood or particle board. Their tubes are often made of cardboard Sonotube™ concrete form tubing. These "cheap" construction materials are not as bad as they sound—wood and Sonotube™ provide a sturdy and thermally stable body for a Newtonian. Unfortunately, the simple Dobsonian mounting is not easily equipped with motorized go-to for automatic object location, which is a desirable—highly desirable—feature for the urban astronomer. Dobs *can* be furnished with Digital Setting Circle computers that will help guide you to objects, but in my experience these DSC systems are not nearly as accurate as go-to. Tracking is also not easily implemented. With a Dobsonian, you'll be nudging the scope continually to keep objects in view, and this can be annoying at the higher powers used in the city.

Thanks to the Taiwanese and Mainland Chinese telescope factories, nicely priced motorized equatorially mounted Newtonians that can automatically track the stars are now available in addition to alt-azimuth mounted Dobsonians. The equatorial telescopes are generally found in smaller focal ratios than the dobs, something that, as we've said, the urban observer may not find as useful as higher focal ratios. But the high quality of the current Chinese mirrors means that the equatorial scopes can often be pushed to as high a magnification as needed while maintaining sharp, clear images. The larger apertures of these scopes, 8 and 10 inches, means that even at $f/5$ it's fairly easy to produce high powers without intolerably short eyepiece focal lengths.

All in all, Newtonians are a laudable choice for the city. They fulfill our prime requirement—they deliver lots of light—at bargain prices. They are also easy to transport in sizes up to about 12 inches, don't require long to adjust to outdoor temperatures—any telescope mirror must cool down or warm up to ambient outside temperature before it can deliver its best images—and can provide images easily equal to those of any other design of telescope.

One thing to be aware of when considering a Newtonian as a city scope is what I call "The Only Enemy of Good Enough is More Better" syndrome. With prices for Dobsonian Newtonians so reasonable, novices may be tempted to buy bigger than is easily portable in an urban environment. Consider 12 inches the *absolute* upper limit and 10 inches the *practical* aperture limit for most city-based observers.

Meade Dobsonians

Meade Instruments has been selling a line of Sonotube Dobsonians (Plate 3) since the early 1990s, and these simple telescopes have garnered much praise for their optics. Until recently, these "StarFinder" Dobsonians were available in 6-, 8-, 10-, 12-, and 16-inch apertures. Meade has now changed focus, going to Schmidt Newtonians (see the "Catadioptric" section) for their smaller non-SCT design scopes, and only the 12- and 16-inch StarFinders are now available new. The 6-, 8-, and 10-inch sizes are plentiful used, though.

While all these telescopes have surprisingly excellent optics, you'll often hear them described as "kits." That's because at the very low prices Meade asks for these scopes (845 U.S.$ for the 12.5-inch $f/4.8$, and 1,245 U.S.$ for the 16-inch $f/4.5$) they've had to scrimp on everything other than the mirrors. The focusers (plastic) and finders

(30-mm) need immediate replacement, and the nylon bearing pads should be discarded and Teflon used in their place if the motions of these scopes are to be smooth.

Despite this kit status, the 12.5-inch scope, especially, can be a real workhorse for the urban astronomer. It's a little heavy, true, but if you can observe from your backyard it's not bad to set up at all, though younger and lightly built observers should, if possible, try moving the 12.5-inch scope unassisted before committing to it. The generous aperture allows you to cut through light-pollution to the point where *galaxy after galaxy* in Virgo, for example, is visible from many sites. Globular clusters take on form and substance as great globes of stars and not just round, fuzzy balls with this scope, and, should you want to look at the planets, the 12.5-inch StarFinder can provide surprising results, besting some much more expensive competitors. The 16-inch? Unless you've got a situation where you can rig up a means of rolling this scope out of a garage or shed to observe, it's really too big to consider. Also, the inexpensive particleboard mount that works reasonably well with the 12.5-inch StarFinder becomes a weak point (literally) with this much larger and heavier instrument.

Synta

These Chinese Dobsonians are, frankly, tremendous buys, and will find favor with any amateur. The optics, while maybe not quite as good as those found in the Meades, are very good indeed, and the attractive metal tubes, nicely finished rocker boxes, and good assortment of included accessories (depending on the seller, often two Plossl eyepieces, a simple collimation tool, and a spring system to help with tube balance) make these imports attractive for everyone. Their rugged construction and their ability to be transported in one-piece (the spring-aided balance system holds the tube and rocker box mount together if desired) makes these nearly perfect for the urban astronomer. The Synta Dobsonian lineup at this time includes an $f/8$ 4.5-inch, an $f/8$ 6-inch, an $f/6$ 8-inch, an $f/4.7$ 10-inch, and an $f/4.9$ 12-inch. Like Synta's refractors, these Dobsonians are available from a variety of distributors in the U.S. and Europe under various brand names and with varying prices.

My only negative comment regarding these telescopes is that the metal tubes, while attractive, actually have poorer thermal characteristics than the cardboard Sonotube™ found on other inexpensive Dobsonians—it may take longer for them to adjust to outdoor temperatures than telescopes with cardboard tubes. The Synta dobs are real winners, though, in the city or out, and the 8-inch, in particular, provides a wonderful combination of performance, price, and focal length for use under light-polluted skies.

Premium Dobsonians

"Dobsonian" doesn't just mean "cheap" anymore. Amateurs can now purchase premium priced and premium quality near-custom-made dobs from companies like Obsession and Starmaster in the U.S. These telescopes, which can be had in apertures of 30 inches and *larger*, are only sold in truss tube configurations. Although their prices are high, large truss tube dobs can deliver remarkable performance to the user who is able to store and transport them.

The truss tube telescope design (see Plate 4) is the current darling of Dobsonian users. In addition to a rocker box mount like those of the Sonotube dobs, it consists of a mirror box that holds the primary mirror, truss poles that take the place of a solid tube, and an upper cage assembly that carries the secondary mirror and focuser. The purpose of this design is to make large-aperture Dobsonians practical—these big scopes can be easily broken down into manageable components for transport. Keep in mind, though, that even the mirror box of a large truss tube telescope will be impossible for many apartment dwellers to manage. Not only will you have to get it and the other components down to street level, something that may be very difficult in apertures above 16 inches, you'll have to find a place to keep it. I still think 12.5-inch scope is a good maximum aperture for most city dwellers, no matter what the scope design.

Europe and the U.K.

Until fairly recently, the Dobsonian has not enjoyed as much popularity in the U.K. and in Europe as it has in the U.S., but that is now changing. In addition to native makers of custom, high-quality truss tube scopes, Meade and other U.S. dobs are regularly imported, as are the Synta Dobsonians, which are sold under the Skywatcher, Konus, and other labels.

One interesting entry in the U.K. is the imported Helios "Skyliner" Dobsonian. This reasonably priced scope, in addition to the features common on other far-eastern dobs, includes alt-azimuth setting circles, which, in conjunction with a suitable computer program, can help users find objects. This is a nice feature for urban observers, and one I'm surprised other dob makers have not picked-up on.

Equatorially Mounted Newtonians

There was a time when it looked as if Newtonian reflectors mounted on German equatorial mountings (GEMs) were as an endangered species as the refractor. Only a few were found for sale, and many of these were in the premium-price category. This style of telescope hadn't changed with the times, either. Heavy (but not overly steady) mounts were the order of the day and discouraged amateurs who needed to move the scope around the yard or down flights of stairs from considering one.

Things began to change for the better when Meade brought out a line of GEM Newtonians in the 1990s, the StarFinder equatorials, which were somewhat successful despite pedestal style tripods that made them less portable than the average city dweller might have wished. Meade built 6-, 8-, 10-, and 16-inch GEM versions (the optical tubes are almost identical to those on the dob scopes) until fairly recently when most of the StarFinder GEMs were phased out with the coming of the LXD 75 Schmidt Newtonian line. At this time, only the 16-inch GEM StarFinder is still available, and it is even harder to transport and shakier on its mount that the Dobsonian version.

Even though most of the Meades are gone, things continue to look better and better for the Newtonian fancier who wants a scope on a mounting that can be equipped with drives for tracking. Synta and another firm in Taiwan, Guan Sheng, are pumping new life into a scope genre that most thought was past its heyday. Orion in the U.S. and Skywatcher and Konus in the U.K. and Europe are, again, the most common names you'll find on Chinese GEM Newtonians.

What makes the imported GEM Newtonians workable is the ubiquitous EQ4 mounting as shown in Plate 5. This telescope mount started life a few years ago as a not-too-good clone of the famous (and now relatively expensive) Japanese Vixen Great Polaris Mount. Early on, the EQ4 couldn't hold a candle to the Vixen, but, somewhat surprisingly, Synta has continued to improve this mounting with advances like the addition of ball-bearings to the RA axis and a better grade of lubricant than the initial glue-like grease found in the first examples to hit the West. While this mounting can sometimes stand disassembly and relubricating by mechanically competent amateurs, it works quite well off the shelf now, being almost as good for the visual observer as the Vixen. Its low price (the EQ4 is available sans drives for less than 300 US$) means that inexpensive—but capable—equatorial Newtonians are now possible.

Synta's GEMs

While equatorial Newtonians in apertures less than 6 inches are plentiful, my opinion is that the urban deep sky observer should pass these by. The 6-inch Syntas are so inexpensive (less than 400 US$) that there's really no reason to settle for smaller aperture. Up to this time, the most widely available Synta scopes have been a 6-inch $f/5$, often seen on the EQ3 mount, which is sufficient for the 6-inch, but noticeably lighter than the EQ4, and an 8-inch $f/5$ on the EQ4 mount. These scopes are capable of good work, and though, as was mentioned earlier, the urban user might wish for a little extra focal length to make higher magnifications possible without having to resort to shorter focal length eyepieces or barlow lenses, at $f/5$ these tubes don't stress their mounts and they will easily take higher powers.

Orion U.K. and Parks (U.S.)

Don't want an inexpensive Chinese equatorial Newtonian? There are premium GEM Newtonian models available even in this era of big dob and SCT domination. In the U.K., Orion Optics (no relation to the U.S. firm of the same name) offers *very* high quality 6–12-inch tube assemblies mounted on Japanese made Vixen equatorial mounts. The smaller apertures on their Vixen Great Polaris mounts are fairly easy to haul around despite the fact that piers are provided as standard equipment in lieu of tripods. In the U.S., the California firm, Parks Optical, goes all the way up to 16 inches with Newtonians on in-house-built mounts. While the optics are usually outstanding, the mountings on the Parks scopes could have been brought straight from the 1960s in a time machine. Very heavy and somewhat shaky, they are not something I'd want to handle in the city.

GEMs to Avoid

One type of equatorially mounted reflector to avoid is the "short tube." Short tube refractors can be nice, but most of the short tube reflectors just don't make it. A "short tube" reflector can be easily identified by its specs. The given focal ratio ($f/10$ is common) or focal length of the scope is much larger than the short, stubby tube of the scope would indicate. The manufacturers achieve this by installing a built-in barlow amplifying lens in the light path. This barlow does extend the focal length of the fast Newtonian mirrors used in these telescopes, but the quality of this barlow (which is all it is, though advertising often refers to it as a "corrective optics assembly" or similar) is usually low, with the result often being excess color and soft images.

Stay completely away from the Chinese GEM-mounted 6-inch Newtonians found on Ebay. These telescopes possess spherical-shaped mirrors rather than the parabolic primary mirrors required for sharp images in a Newtonian. This results in *extremely* poor images at the low focal ratios of these scopes' primary mirrors ($f/5$ or $f/4$ usually).

Catadioptric Telescopes (CAT)

What would be the characteristics of the perfect telescope for exploring deep space from the city? It would have a long focal length that would allow us to achieve medium and high powers easily. It would be easily adaptable for DSCs or go-to, making it easy to find elusive objects in guide-star deprived urban skies. It would feature generous aperture to pull faint fuzzies out of the murky city heavens. Above all, it would be compact and portable. For many amateurs the telescope design that fulfills these requirements best is the catadioptric (CAT), the lens–mirror hybrid that has become exceptionally popular with amateur astronomers over the past 30 years.

Schmidt Cassegrain Telescope (SCT)

The SCT, the Schmidt Cassegrain telescope (Plate 6), is not the *only* CAT used by amateurs. The Maksutov Cassegrain, the Maksutov Newtonian, and the Schmidt Newtonian have their fans, too. However, while I might be accused of being a little biased, since I've become known as "Mr. SCT" for my unabashed love of this telescope design, there are good reasons why the Schmidt Cassegrain telescope may be the ultimate "city scope." There's a lot that can be said about SCTs, you could even write an entire book on the subject (I have), but "SCT theory" can be summed up in relatively few words.

Being a catadioptric scope, the SCT includes a lens in its design. This is the large, thin corrector "plate" mounted at the end of the tube. SCTs use easy to fabricate spherical primary mirrors. Unfortunately, in a condition similar to chromatic aberration in refractors, a spherical mirror can't bring all rays of light to the same focus. This would result in blurred images lacking in detail. The SCT's corrector plate removes this spherical aberration and allows its spherical mirror to produce images as sharp as a Newtonian's parabolic-shaped primary.

Meade and Celestron SCTs

Want an SCT? I don't blame you. Luckily, the choices here are fairly easy since the telescopes made by the California firms Meade and Celestron are the only ones available outside very expensive semi-custom units produced by a few firms in limited numbers. What are today's SCTs like? If there's any telescope make that has embraced fancy computers and electronics, it's the Meade and Celestron SCTs.

Current top of the line scopes, Meade's LX200 GPS series and Celestron's Nexstar GPS line, feature a bevy of electronic options including automatic alignment via built in Global Positioning System satellite receivers, digital compasses and electronic levels. These scopes will, when powered up, find their position, locate North, and slew to appropriate alignment stars. All the user has to do is center these stars in the eyepiece as instructed by the display on the telescope's computer hand-controller and hit "Enter." Thereafter, both the Meade and Celestron GPS SCTs will reliably and accurately point at any object in the sky.

The Meades are available in apertures of 8, 10, 12, 14 and 16 inches while the Celestrons come in 8, 9.25, 11 and 14-inch sizes, the prices of which begin at 2,195 U.S.$ for the Meade and 1,999 U.S.$ for the Celestron. All Meade and Celestron SCTs are $f/10$ focal ratio scopes, Meade having recently discontinued its $f/6.3$ optics option. Celestron also sells a line of GPS compatible go-to-equipped heavy duty GEM scopes, the CGEs, but their large, heavy mounts are really a bit much for urban observers.

The next tier down in SCT-land is non-GPS go-to scopes. These models, Meade's LX90 and LXD75 SCTs and Celestron's Nexstar(i) and AS series, feature full go-to capabilities, but the user must set time, date, latitude and longitude; and find North before the scopes will locate objects on their own. While not as convenient as the GPS units, spending a few minutes entering data into the scope hand controller is a small price to pay for being able to view any one of the thousands of objects in these telescopes' internal libraries at the touch of a button for the rest of the night. Meade's LX90 (8-inch) and the Celestron Nexstar (i) (5 or 8-inch apertures available) are traditional fork-mounted telescopes, while the new LXD75 SC-8-inch SCT and the Celestron AS GT scopes (available in 5, 8 and 9.25 and 11-inch apertures) use Chinese German equatorial mounts equipped with computer controlled drives.

The Meade LXD 75 SC8 SCT uses the same Chinese GEM mounting as the AR5 and AR6 refractors, and, like them, and seems fairly accurate when correctly aligned. Enhanced coatings, the UHTC group, are available for this telescope and for all the SCTs Meade makes for an extra charge.

The Celestron AS GT telescopes use a go-to-equipped Synta EQ4 mount (called the "CG5" by Celestron) that works amazingly well considering the fairly low cost of the GT series. All these telescopes are available with Celestron's version of enhanced coatings, "XLT," for slightly more money and with non-computerized, non-go-to EQ4 mounts for less.

The Celestron fork-mount go-to scopes, the Nexstar 8 and 5(i) possess single-arm forks rather than the usual two-arm design seen on the Nexstar GPS scopes and all the Meades, but they are quite stable and also very light and easy to transport. One interesting feature of these Celestrons is that the user can add an external GPS receiver module to them if desired, giving them many (though not all) the capabilities of the GPS scopes. The Nexstar 5 offers decent aperture and go-to in a lightweight package. This is the only sub-8-inch SCT currently available.

The non-GPS Meade fork-mounted telescope, the LX90, is fully computerized, and is one of the most popular and most reasonably priced go-to telescopes made. For a surprisingly low price, the user gets a computer equipped instrument capable of being used in comfortable, stable alt-azimuth mode attached directly to a tripod, or in equatorial configuration on an optional "wedge," where it is capable of taking amazingly good long exposure deep sky photos.

At the bottom of the SCT pyramid are the non-go-to manually operated SCTs. Unfortunately, these appear to be a dying breed. Meade and Celestron are of the opinion that consumers want computerized telescopes now, so offerings in this realm are limited. Meade has just discontinued its only remaining manual fork-mounted scope, the 8-inch LX10. Celestron has likewise discontinued its non-go-to Celestar.

At the time I'm writing this, it's still possible to get a new LX10 from many dealers, and that might not be a bad idea if you want a solid, inexpensive SCT that's not loaded down with computers. This is a "traditional" fork-mounted SCT quite similar to the first models Meade offered over 20 years ago. It is a nice little scope with a modest price, a light but reasonably stable fork mount, and an accurate motor drive powered by a 9-volt battery.

Again, if I had to choose a telescope design for city-bound observing, it would be the SCT. No doubt about that. It's just so darned adaptable: Lots of focal length with the ability to "reduce" it for wider field views using optional "reducer/corrector" lenses. Easy adaptability for film or CCD imaging. Fully computerized go-to, which is a wonderful, almost indispensable tool for the urban astronomer. Relatively good transportability in apertures from 5 inches to 11 inches. If there's not a perfect scope for the urban scene, the SCT at least comes close.

While SCTs are considerably more expensive than Newtonians, they are still surprisingly inexpensive considering their many features. This leads some city observers to decide to forego an 8 or 10-inch SCT in favor of a 12- or 14-inch one. Think long and hard before considering one of the "big ones." While a GEM mounted 14-inch like the Celestron CGE 1400 might be barely practical in the city, one of the Meade 12- or 14-inch fork-mounted scopes is big and tremendously heavy, much bigger and heavier than any apartment dweller will want to own. Be aware that the 14 inchers, especially, are much larger than they look in those pretty magazine advertisements. They are also far too heavy for one person to lift, much less carry down several flights of stairs.

The SCT shopper should be aware that models come and go rapidly in the competitive SCT market, with Celestron currently being in the process of introducing a lower priced GPS scope, the "CPC," and Meade coming out with a line of premium SCT-like instruments, the RCX-400s.

Maksutov Cassegrains Telescope (MCT)

The Maksutov Cassegrain, "MCT," an example of which, the Meade ETX 125, is shown in Plate 7, is very similar to the SCT. The only differences are the thick, spherical "deep-dish style corrector plate" and larger focal ratios of the MCTs. You often hear beginning amateurs being advised to "stay away" from MCTs because their large focal ratios, $f/12$

to $f/15$ usually, and resultantly longer focal lengths "make them unsuitable for deep sky work." Nothing could be farther from the truth. All the large focal ratios mean is that the MCT owner must use longer focal length eyepieces to reach low magnifications than an SCT user, for example. While this can create problems for the suburban or country astronomer trying to squeeze every last of bit field size out of her scope, it is really not a concern in the city. Urban astronomers *want* more magnification. Actually, even under dark skies, many DSOs can benefit from more power than most observers apply. And anyone, anywhere can benefit from the usually superb optics of MCTs.

While the MCT is, for some observers, saddled with longer focal lengths, it pulls ahead of the SCT—at least a bit—in optical quality. The sphere-shaped corrector plate/lens that is used in the Maksutov Cassegrain design is decidedly easier to fabricate than the SCT's thin, complex-shaped corrector, so the MCT corrector is easier to do well in all apertures. Because of its longer focal length main mirror, the secondary mirror on MCTs is smaller than that on SCTs, enabling the Maksutovs to deliver slightly more contrasty images than the SCT, images often on a par with those delivered by APO refractors costing many times more. Drawbacks? Not many. The biggest is that MCTs above about 7 inches are very expensive. That is because the corrector must be made from a single, very thick piece of glass. Finding an adequate "blank" becomes an expensive proposition in the larger apertures.

The Meade MCTs

If one MCT has come to mean "Maksutov" in the minds of today's amateur, it's Meade's amazing ETX. This remarkable scope began as a 90-mm clone of the very expensive Questar MCT, but it has evolved over the years into a little wonder equipped with a go-to drive system (using the same Autostar hand controller supplied with the LX90 and the LXD75) and is now available in apertures of 90, 105, and 125-mm. These instruments provide very fine images on a par with any telescope in their aperture ranges.

Liabilities? A few. The go-to system can be slightly difficult for beginners, especially those without much knowledge of computers, but this goes for any go-to-equipped scope. The ETX does use a lot of plastic in its construction to keep both cost and weight down. In most cases this has not caused problems with the scopes, and Meade continues to improve both ETX hardware and software. If there's any real problem with the ETXes, it's their aperture. The maximum available, 5 inches, isn't always adequate if your sky is as bad as mine is. But a light, inexpensive go-to scope like this is a godsend for many urban users. All the ETX models, like Meade's SCTs, can be ordered with enhanced UHTC coatings.

Meade also offers a larger MCT, the LX200 GPS 7-inch Maksutov. This Mak, on the same go-to fork-mount system offered for the SCTs, is full featured and is and currently sells for 2,695 US$, which is less than the price of many manufacturers' 7-inch MCT optical tubes alone. The 7 has, unfortunately, been a problematical scope over its lifetime. One difficulty has been the optics: the 7-inch Maks have varied from barely average to excellent from sample to sample. It now appears that Meade has cleaned up its optical act with this instrument, however. Then there's cool down. Contrary to popular belief, most MCTs do not have inherently longer cool down times than SCTs, but the Meade 7 is a different animal.

Like most MCTs, the LX200 Mak has a longer tube than an SCT, aperture for aperture. This is due to the longer "native" focal length primary mirror used in this design. The 7-inch with its longer tube and heavy corrector plate was difficult for

Meade to balance on the fork mount. To remedy this, *a lead weight* was added inside the scope's rear cell. This worked, but, unfortunately, the lead soaks up a lot of heat and radiates this for quite a while, meaning cool-down time is extended. Any telescope needs to adjust to outdoor temperatures before its optics can deliver their best images. A built in fan helps with cool down to some extent, but the scope still needs quite some time to cool off compared to an 8-inch SCT. Despite these difficulties, the combination of a "big" MCT and the LX200 GPS mount is a powerful one for city-dwellers.

Russian MCTs

Since the early 1990s, Russian MCTs from the firms Intes, Intes Micro and Lomo have been aggressively marketed in the West. In the beginning, quality was catch-as-catch-can, but the past ten years Russian output has improved to the point where these telescopes can give even the highest-toned APO refractors some pretty stiff competition. Of particular interest are the Intes 6-inch MK67 and MK66 MCTs. These Maksutov Cassegrains are optically excellent and provide enough aperture for the urban astronomer to actually see something. The only difference between these two models is their focusing systems. The MK67 focuses via a Crayford focuser (which moves the eyepiece in and out to focus). The MK66 is intriguing since it uses a moving mirror focuser and accepts any of the many accessories developed for American SCTs. Unfortunately, Intes has recently announced that it is leaving the amateur telescope business, but the 66 and 67 should remain available new for quite some time.

When it comes to mountings, most Russian scopes are sold either without mounts or on Chinese GEM equatorials. Lomo's MCTs, which are just now becoming popular with Western observers, are available in a fork-mount configuration that looks for all the world like a 6-inch SCT. One mistake that many buyers of these relatively inexpensive MCTs make is skimping on the mounts. Maks are almost always heavier than SCTs, and when this is combined with their large focal ratio high magnification characteristics, they demand steady support. A Synta EQ4 or a Vixen Great Polaris is the minimum for a 6-inch Maksutov Cassegrain.

Orion U.K.

One of the most interesting and reasonably priced new MCTs to become available recently is the OMC 140 from Orion Optics in the U.K. This scope is competitive with the Russian instruments in every way, and, from early reports I'm hearing, it may be a notch above the Russians in optical quality. The telescopes are available with a selection of GEM mounts including the EQ4, Great Polaris, and Vixen GPDX German mounts.

Premium MCTs

Questar

If an humble (but optically good) ETX or Intes does not appeal to your sense of style, there are premium priced MCTs available that have for years defined "luxury" in the amateur astronomy world. First and foremost are the Questars. These MCTs, available in 3.5-inch (Plate 8) and 7-inch apertures, were the first Maksutov Cassegrains to become popular with amateurs, and have remained popular despite their always-high prices. The 3.5-inch fetches an amazing 4,000 US$ with minimal accessories, and the new Titanium-bodied fork-mounted 7-inch pushes even higher at 9,000 US$. These are beautiful telescopes, of course, but 3.5 inches is 3.5 inches and 7 inches is 7 inches. Even the very expensive Q7 is a little weak aperturewise when you have to battle heavy light-pollution.

Telescope Engineering Company (TEC)

A little less expensive and larger, too, are the TEC MCTs. This U.S. company, which has made a name for itself with high-quality optics, sells some astoundingly good MCTs. While expensive compared to Meade or Lomo, they are fairly affordable considering their large apertures (for Maks). The TECs range from an 8-inch at 3,700 US$ to a 12-inch at 12,600 US$—these figures are for optical tubes only. At a focal ratio of $f/11$, the 8-inch is both compact and blessed with a comparatively short tube length—this could be just the telescope for the observer wanting a "super SCT." TEC does not provide mountings for its scopes, the expectation being that the buyer will want to choose a quality GEM appropriate for these fine instruments from a third party.

Astro-Physics (AP)

Roland Christen's AP, while noted for its APO refractors, has entered the catadioptric arena recently with a 10-inch MCT that is worthy of the AP name. As with all of this company's other products, the 10-inch is built to the most exacting standards, is produced in limited numbers (meaning that you'll be on a long waiting list for this one just as you would be for one of the refractors), and commands a price commensurate with its quality, 10,900 US$ at the time of this writing. This scope is, like the TECs, sold as an OTA only, but AP can furnish its high-tech go-to GEMs for use with this Mak, the big 900 or 1,200 mounts being suggested for this sizable OTA. For the urban astronomer not fazed by the near 20,000 dollar price tag once a top of the line 10-inch system has been assembled, the only "minus" consideration is the considerable weight of this telescope. The 10-inch OTA is fairly heavy at 33 pounds. Add to this a mounting appropriate for this big telescope and you wind up with something you won't be moving around the backyard in search of tree-free zones.

Schmidt Newtonians Telescope (SNT)

The Schmidt Newtonian design is simple and elegant. Its primary optical component is a spherical primary mirror like those used in SCTs and MCTs. The main difference is that there's no central hole in this mirror for the eyepiece to "look through." As in a Newtonian, light collected by the primary mirror is focused back up the tube to a flat secondary mirror angled at 45° that sends the light through a hole in the upper end of the tube to a focuser. This design incorporates an SCT-style corrector plate. As in the SCTs and MCTs, the corrector lens allows the scope to use the simple spherical primary mirror without the penalty of spherical aberration. It also enables the scope to produce a wide field that's flatter than that delivered by Newtonians. The practical result is that, f-ratio for f-ratio, the SNTs produce much sharper star images out toward the edge of the field than standard Newtonian reflectors. In an $f/4$ Newtonian, stars on the field edge are far from sharp, resembling little comets or seagulls more than pin-point stars. In an $f/4$ SNT, they are satisfying pinpoints.

The Meade SNTs

The SNT has long been neglected by the major telescope makers, with only Meade choosing to produce these telescopes in any numbers for any period of time. Even Meade abandoned the design for a while. No commercial SNT had been available for years when Meade revived the design in 2002 for the LXD55 series. The LXD55 SNTs and the recent upgraded replacement for them, the LXD 75 series, are a 6-inch $f/5$, an 8-inch $f/4$, and a 10-inch $f/4$. These telescopes, mounted on the same computerized GEMs used for the aforementioned AR-5, AR-6, and SC 8 telescopes, and are very inexpensively priced.

I've been able to personally test the 10-inch SNT and was rather impressed. Tube fittings and the scope in general seemed of higher quality than the minuscule price would suggest. Optics were good and sharp, producing better images than a comparably priced Newtonian, and seemed equally capable of wide fields and high powers. My only reservation is of the "why" variety. Most of the time, we city dwellers can't make much use of the relatively low powers these scopes can achieve, and would be happier with a plain, old SCT. The SCT, in addition to it's larger focal ratio, also has in its favor the fact that it can use the hundreds of accessories available for Schmidt Cassegrains—the SNT is an uncommon and rather specialized scope.

Maksutov Newtonian Telescope (MNT)

If you understand the design of the Schmidt Newtonian, the layout of the Maksutov Newtonian is easy to grasp. Take an SNT, make the focal ratio of the primary mirror a little larger ($f/6$ is common), remove the Schmidt corrector plate, replace it with a Maksutov corrector, and you're done! Like MCTs, the MNTs generally stick with longer focal lengths than their Schmidt brothers. This has the advantage of allowing the use of a smaller secondary mirror, which helps improve contrast. If your goal is refractor, APO refractor, quality images, larger aperture, and smaller prices, the MNT may be your telescope, as these are its strengths.

The Intes MNTs

While a Russian company with a similar name, Intes Micro, is also a producer of MNTs, most of the current enthusiasm for these scopes has been generated by the two superb instruments made by Intes. As mentioned earlier, Intes is leaving the scope business, but the MNTs were produced in numbers and will be available for a while new and indefinitely as used scopes. The Intes MNTs include the 6-inch $f/6$ MN61 and the 7-inch $f/6$ MN71. Both telescopes are equipped with secondary mirrors that are small for their apertures compared to those in SCTs or even Newtonians—30 and 36-mm, respectively. This, coupled with the forgiving Maksutov design and superb quality optics in general, makes these scopes come very close to equaling APO refractors of equal aperture. The medium focal ratios of these instruments help them provide both low powers for wide field use and high powers for planets and small DSOs. Their ability to handle higher powers adroitly and good contrast characteristics make these scopes naturals for the city. About the only nits I can pick with regard to these instruments is that their apertures are small for their prices compared to Newtonians, and are the significant weight of these scopes reduces their portability. The 6-inch is manageable at 20 pounds, but will require at least an EQ4 mount. The 7-inch model's 31 pounds will make a mount in the next weight (and price) class a must.

Urban Accessories

What have you got if you've got a scope without accessories—eyepieces, filters, and all the other little doo-dads active observers accumulate? A big paperweight. Every new astronomer will soon come to the realization that astro-buying has just begun with the acquisition of a telescope. Accessories are necessary for everyone, but the choice of proper eyepieces, filters, and other necessary items is particularly critical if you're observing under city lights.

CHAPTER THREE

Accessories for Urban Observers

Accessories are the bane and delight of every amateur astronomer's existence. You never stop buying. Of course, many of the "must-haves" are really superfluous impulse purchases. Who can resist buying *just one more* eyepiece from a dealer's table at a star party? But there is no doubt that you really do need quite a bit of equipment in addition to a telescope in order to enjoy productive observing.

Many of the accessories you'll need as an urban observer are exactly the same as those required by any working amateur astronomer. But some are more important for us than they are for country observers—light-pollution filters, for example. Sometimes one type of accessory is better for city use than another. We tend, for example, to gravitate toward medium focal length eyepieces for "low power" use rather than the long-focal length "big glass" used under dark skies. In this chapter, we'll discuss what you need and how you should invest those usually limited astronomy dollars wisely.

Light-Pollution Reduction Filters

When a new city lights astronomer starts seeing advertisements for "light-pollution" or "light-pollution reduction" (LPR) filters, she or he immediately thinks the problems with bright city skies are over. Screw one of these filters (Plate 9) onto the end of an eyepiece and you're magically transported to dark country skies and galaxies are everywhere! Sorry, but it just isn't so. LPR filters can help, but, as always, there ain't no such thing as a free lunch. Filters are a compromise, and cannot substitute for dark country conditions, but they can make your city-based viewing more profitable and more enjoyable.

LPR filters for use by amateur astronomers started appearing in the 1980s, coincidental with expanding suburbs and growing light pollution in the U.S. and Western Europe. The concept behind these filters is simple: coat a piece of optically flat glass with multiple layers of substances designed to reflect certain wavelengths of light. These coatings are chosen so that all wavelengths generated by Earthly light sources—incandescent, sodium, and mercury vapor lights—are rejected by the filter, reflected away by the coating layers before they enter the eyepiece. The LPR filter does not make deep sky objects (DSOs) brighter. Its ability to reject light-pollution wavelengths merely means that it increases the *contrast* between the sky background and the object of interest. The background sky becomes darker due to the subtraction of earthly sky glow without the target object being made much dimmer by the filter.

With one of these LPR filters screwed onto the telescope end of your eyepiece, then, only the light from distant DSOs reaches your eyes? Dark sky paradise in downtown London or Manhattan? Unfortunately, this good idea doesn't work quite as well as we'd wish. Like anything else, an LPR filter is not 100% effective. Some unwanted wavelengths do make it through the filter. The main problem, though, is that the light of the stars falls into the same range, the same band of wavelengths, as that from man-made light sources. That means that in addition to blocking the light from the ground, your filter also blocks the light from the stars. Because of this, we can immediately eliminate light-pollution filters as a tool for viewing galaxies. Galaxies are composed of stars, and *their light is attenuated by filters.* Forget other stellar subjects—open and globular star clusters—too.

So why waste your money on an LPR filter (they sell for about 100 US$ or 100 UKP)? What good are they? They won't do a thing for clusters and galaxies, but LPR filters can work spectacularly well for diffuse and planetary nebulae. This is a good thing, since nebulae are probably hurt more by light pollution than any other object, even more than galaxies. The presence of an LPR filter can easily make the difference between getting a good view of a nebula from the city and not seeing it at all. But *which* filter? Dealers' advertisements tout a number of oddly and confusingly named products, "OIII" ("oh-three"), "UHC," "hbeta" and more.

At this time, light-pollution filters, no matter who makes them, fall into three categories: "mild/broadband," "medium/narrowband," and "line filters." Mild filters are represented by the Meade Broadband and the Orion U.S. Skyglow, and are available from a number of other manufacturers as well, both in the U.S. and Europe. Their primary characteristic is their wide "passband." They pass a wide range of wavelengths of light. That is both good and bad. It's good in that these filters will "work" on many different objects, and don't dim field stars much. The "bad" is that they don't reduce background sky glow much, either. By allowing-in a fairly wide range of wavelengths, including those emitted by the stars, a large amount of earth-based light sources are also passed.

Should you buy a mild filter? In my judgment, probably not. They do very little for the visual observer. Sure, the field looks good, with lots of stars visible. Even galaxies are not dimmed much by a wideband. Unfortunately, neither is the background glow of light pollution. These filters darken the field slightly, and only slightly. They may improve the view of a nebula that's easy to see from the city already—M42 in Orion, for example—but only minimally. They are mainly of use for photographers who need or want to try to image with low focal ratio scopes from the city, or by country observers seeking to enhance views of nebulae a little without attenuating field stars.

I've occasionally heard broadband filters referred to as "galaxy filters." It would be nice if there were a light-pollution filter that could enhance galaxies, of course, and some wideband users claim that they darken the sky enough without dimming the target galaxy to improve its appearance. My tests reveal, however, that, without exception, galaxies look better without one of these filters than with one in or out of the city. The broadband filters are not galaxy filters. Sadly, *there are no galaxy filters.*

Next up are the narrow-band filters. These filters, like the Orion Ultrablock or the Sirius Optics Nebula Filter, are the bread and butter of the urban astronomer. When properly used, they can make a nebula that is badly compromised in the city, like M17, the Omega Nebula, into a near-showpiece object. I've been constantly surprised at what I can pull out of the sky glow with a narrow-band. Not just the Messier nebulae, either; many faint and obscure NGC and IC clouds are routinely visible from my city sites with an 8-inch SCT and one of these filters. The faint nebulae scattered through Cygnus, Cepheus, and Cassiopeia don't exactly become spectacular—again, LPR filters are *not* a substitute for dark skies—but they are detectable and even enjoyable.

Narrow-band LPR filters work by allowing only very narrow slices of the visible light spectrum to pass. Most filters of this type are actually designed with *two* passbands, one centered on the hbeta region of the spectrum (the red light of hydrogen emitted by many nebulae) and the OIII area (light from the doubly ionized oxygen atoms often present in planetary nebulae). The rest of the spectrum, emitted by mercury vapor and other man made sources, is attenuated to a surprising degree.

Narrow-band filters are wonderful tools for the urban observer. The only thing you may not like about them is those relatively narrow passbands. With these filters, you get to the point where the stars are being obviously dimmed, and you may find that the star fields don't looks as pretty with the filter in place as without. In my opinion that is a small price to pay for actually being able to *observe* dim nebulae from home. As to which brand to choose, all the narrowband filters on the market in the U.S., UK, and Europe are quite similar, with the main differences being the comparative width of the passbands in the OIII and hbeta regions. Some let in more red hbeta light and others more green OIII (don't expect to see color differences on dim nebulae visually, of course). Practically speaking, the performance of all the narrowband filters I've tried has been remarkably similar. Some current manufacturers of these filters in addition to Orion and Sirius are Baader Planetarium, Thousand Oaks, Meade, Celestron, and TeleVue. All are widely distributed throughout the world.

The third type of light-pollution filter is the "line" filter, which is widely available in two varieties—the OIII which, as you'd guess, has a passband centered on OIII light, and the hbeta, which concentrates on the red section of the spectrum. These filters take their name, "line filters," from the fact that their passbands are so narrow that they are designed to admit only the light from the emission lines of hydrogen or oxygen.

After having been involved in amateur astronomy for 40 years, I've spent a lot of money on accessories of all types—all of which were described in glowing terms by their manufacturers. Only occasionally have I found a piece of equipment that has really lived up to my expectations. The OIII filter is one of these; it is a truly remarkable filter. The OIII can take the dimmest planetary nebulae and make them come alive for the urban observer. With it, I've been able to at least *detect* the Helix Nebula in Aquarius—usually considered a dark sky object only. The Veil Nebula in Cygnus also shines out (dimly) on good nights. Many other planetary nebulae also take on more form and substance with the application of the remarkable OIII.

Nothing in this world being perfect, the OIII isn't all gravy. This is a very *strong* filter, and will definitely subtract dimmer field stars from your view. In addition to making the field less attractive, this can also make it harder to get your eyepiece in focus. The addition of an LPR filter to your eyepiece changes its focus—you have to refocus the scope when you add a filter. This refocusing is easy enough with a broad or narrow-band filter. Just focus on a field star. With the OIII filter in place, however, there may not *be* any field stars to focus on, and it can be quite difficult to get the scope adjusted correctly with only a dim planetary nebula and a few dim stars as subjects. Often you'll find yourself slewing the scope to a bright star that will shine through the OIII. Once focused, though you may have difficulty locating your nebula again.

The OIII tends to work best with larger aperture telescopes in my opinion. To be most effective, the OIII needs a fair amount of light, it seems. This is not to say that it can't be used at all with smaller scopes. I've had some nice views of brighter and larger DSOs with an OIII used with my 80-mm *f*/5 refractor, but the views of any nebula through the OIII seem better with increasing apertures.

Finally, the OIII does not work on all objects. Not all nebulae or even all planetary nebulae radiate strongly in OIII light. While most OIII filters will admit at least some hbeta radiation, an object must be pretty bright in OIII light for the filter to work well. Most diffuse nebulae are improved by the OIII, but some are not. The Orion Nebula, for example, while not really dimmed by an OIII filter is not significantly improved by it in my opinion, either.

The final commonly available nebula filter for your consideration is the hbeta. This is often referred to as the "California" or "Horsehead" filter. The reason for these monikers is that the hbeta is mainly used in attempts to see those two *extremely* faint nebulae, which radiate almost all their light in the dim, red hbeta band. The hbeta filter is nice if you're chasing these legendary nebulae from dark skies, but is not of much use under sodium streetlight in the city. No matter which filter you use to combat heavy light pollution, you are not likely to catch a glimpse of the Horsehead or California from the city or even the suburbs—even with a rather large telescope.

There are a few other nebulae that respond well to the hbeta, but they are, if anything, even more challenging than the Horsehead and the California. When used on bright nebulae, an hbeta can be somewhat effective, but, in my experience, doesn't improve the view beyond what you'd get with a broadband filter, and certainly doesn't provide the performance of a narrowband type. Red hydrogen light is dim, and shutting off all other sources of light from an object can make almost anything hard to see.

Which to choose? At 100 dollars or pounds for 1.25-inch filters—2-inch units are approximately twice as expensive—these little pieces of glass don't come cheap. In my opinion, a narrowband filter from Orion U.S., Lumicon, or Baader is the place to start. They provide enough "gain"—contrast enhancement—to make a considerable difference in the appearance of nebulae from the city. The narrow-bands also have the advantage of "working" on the largest number of objects. As a second filter, the OIII is a worthwhile investment. Despite the fact that it dims field stars badly, it is an *amazing* filter. If you're interested in planetary nebulae, especially, the OIII belongs in your eyepiece box. The hbeta? If you never observe from dark—and I mean really dark— sites, you can probably do without the hbeta. If you do get to pristine skies occasionally, are interested in chasing the hard nebulae, have at least a medium aperture scope, and possess a lot of patience and observing skill, the hbeta might be nice to buy "some day."

At least as important as *choosing* the correct light-pollution filter is *using* an LPR filter correctly. A few years ago, I began hearing city observers complaining that their light-pollution filters just didn't work at all. According to these folks, not only did their OIIIs and Ultrablocks not *improve* nebulae, they actually made them look *worse*. I knew this wasn't true, as I've used these filters very successfully in the city with scopes as small as 60-mm.

A little investigating revealed the cause of the problem these people were having: that old devil, *ambient light*. An LPR filter is screwed onto the "telescope end" of your eyepiece and works by reflecting unwanted wavelengths of light away before they can enter the eyepiece. What happens, though, if you allow ambient light from man-made sources to enter the *other* end of the eyepiece? From the "eye end?" It enters the eyepiece, hits the filter screwed onto the scope end, and is reflected right back into your eye. Your eyepiece is flooded with light-pollution. That was what was happening to these observers. No wonder their filters made objects look worse. The secret to avoiding this light-flood is to shield the eye lens end of a filtered eyepiece from ambient light. This can be as simple as cupping your hands on each side of the eyepiece or as elaborate as arranging one of the light shields we'll talk about in the following chapter. In addition to helping your expensive LPR work as designed, protecting yourself from ambient light also allows your eyes to attain a little dark adaptation

Finders

Every amateur telescope needs a finder, the small, low-power telescope mounted on the main tube. A big scope has such a small field of view, relatively speaking, that without the wide-angle finderscope, locating objects—even the Moon—is an exercise in frustration. Finders are less important for the users of self-pointing go-to scopes, but the observer must still be able to aim the scope at alignment stars initially, and will need a decent finder to do that. If go-to alignment is a little off, the finder can also help locate an object outside the main scope's eyepiece field, as a surprising number of DSOs are visible in large finders—even in the city.

For the urban astronomer who doesn't use a computerized telescope, an adequate finder is vital. The problem for the city astronomer in locating DSOs is that there are far fewer stars visible in city skies than there are from the country. Out in the dark hinterlands where you can see down to magnitude 6 or better, finding objects is easy. There are plenty of stars in the sky corresponding to those on your charts. Plenty of "waypoints" to help you "star-hop" from sparkler to sparkler till you arrive at your target object. Reduce the number of stars visible to magnitude 4, however, and there will be some areas of the sky that are nearly "blank" to the naked eye. Virgo is a good example. Between the widespread arms of the Virgin, the "Y" shaped western side of the constellation, is the wondrous Realm of the Galaxies. There are dozens of island universes in this area that the urban astronomer can see with a modest telescope. But how do you find them? There very few visible stars spread across the 15 degrees of sky between Vindemiatrix and Omicron Virginis under city skies. The answer is a nice, big finder.

The average finder shipped with a medium-cost telescope has an aperture of 30 mm (I do notice that even some of the less expensive Chinese telescopes are beginning to be shipped with 50-mm finders, lately). This is sufficient, if not generous, for the country. In the city it is almost useless. Get rid of it as soon as you can, as it will not show enough stars to make finding DSOs anything but frustrating. A 50-mm finder, on the other hand, is just about perfect. Even in poor locations, a 50-mm aperture finderscope can deliver stars down to about magnitude 8, meaning you'll be able to see every star plotted on the popular *Sky Atlas 2000* star charts.

Which 50-mm finder? Most of the finders in this aperture range I've seen for sale over the last 5 years have been surprisingly good. Some do produce sharper stars at the edge of the field, but I don't find that overly important for object locating. What's important is a nice wide field—about 4–5° is common and good—and a set of crosshairs that is easy to see and sharply focused. Some observers ask me if they should pay more for an illuminated finder, a finder with a small, red LED attachment that lights up the crosshairs. If you observe from dark sites the illuminator can be nice, as crosshairs tend to disappear under really dark skies. In the city this feature is utterly useless. Our skies are bright enough that crosshairs are easily visible against the sky background.

I've occasionally had calls from disgusted amateurs who've just spent some hard-earned money on a new 50-mm finder (you'll pay around 50 US$ or 50 UKP for an imported 50-mm finder) only to find that it "won't focus." Looking at the current design of 50-mm finders—most finderscopes sold these days are very similar, most being made at the same Chinese factory—there's no obvious way to focus one. The eyepiece is built-in and immobile. Actually, you *can* focus them, although *how* is puzzling since most don't come with instructions of any kind. The secret is unscrewing the objective end. Point the finder at a bright star and unscrew the objective cell a little. You'll find that doing this reveals a locking ring just behind the spot where the objective unscrewed. This can be screwed inward if you need to screw the objective in to achieve sharp focus. But whether you have to screw the objective in or out to focus, you snug this ring up against it to lock it in place when you're finished. Simple and effective.

Are there any 50-mm finders I would rather not have? Yes, non-correct-image right angle finders. Many people don't like looking through a normal "straight-through" finder, which can be a literal pain in the neck. They find one with a built-in 90° star diagonal more comfortable to use. I don't blame them. Craning your neck to look through a straight-through finder when the scope is pointed near the zenith is no fun. The problem is that a normal star diagonal produces an image that is *mirror reversed*. What you see in the finder will never match what's in your star atlas (many computerized star atlases will admittedly allow you to print mirror reversed charts). This makes object location incredibly confusing and frustrating.

If you'd like a right-angle finder, get one advertised as "correct image." These use an amici-type prism to produce views that are not only mirror correct but also right-side-up, making star-hopping a joy. Formerly, one of these finders was an expensive item, but they have recently become available from many dealers in the U.S. and Europe for prices comparable to those of straight through finders. How is that possible? Those ubiquitous Chinese telescope factories again—good quality and rock-bottom prices.

Non-Finder Finders

Over the past 20 years, many observers have begun using zero power sights rather than finder *telescopes* thanks to the genius of the late American amateur astronomer and telescope maker, Steve Kufeld. Steve didn't like the fact that standard finderscopes produce relatively small fields of view and upside-down images. His solution was his Telrad, the "Telescope Reticle Aiming Device" (Plate 10). The principle of operation of the Telrad is similar to that used in the heads-up displays of advanced fighter aircraft: it projects a sighting reticle, a bullseye, into space.

The Telrad is actually a very simple gadget that uses a red LED, a printed transparency, and a couple of batteries to project its reticle onto a piece of glass. When using the Telrad, the observer sights through the glass, and the bullseye reticle seems to float among the stars. For the country observer, the Telrad is a joy. Position the bullseye in the proper place against the stars and you're done. You can purchase clear plastic Telrad overlays for star atlases, and many computer star charting programs will print a Telrad reticle on the maps they produce, making it easy to see where to place the reticle in the sky. This Telrad is a boon for the observer with dark skies, but is not easy for the urban observer to use. I don't recommend one in the city and rarely use a Telrad there myself—not by itself, anyway.

The problem is the lack of guidestars in light-polluted skies. It's hard to place the Telrad bullseye in the right spot among the stars when you can't *see* many stars. Even with a low-power eyepiece in the main scope, you're in for a lot of hunting around. This happens because, unlike a finderscope, a Telrad can't deliver more stars than your eyes alone can reveal—a finderscope's objective easily gathers much more light that your naked eye. The same problem exists for other nontelescopic aiming devices now available for astronomy: not enough stars. If you want to use a Telrad or other zero (or "unit") power finder in the city, fine, but be advised that you'll need to use one as a *supplement* to your finderscope, not as a replacement for it. In combination with a good finderscope, however, a Telrad can really speed up your finding. The Telrad places you in the approximate vicinity of the object, and you can then zero-in on it with the finder without a lot of slewing around. Get a Telrad, but use it with a finder.

Eyepieces

Eyepieces that are good in the country are usually good in the city and vice versa. Conditions in the city *are* different, however, and it's a good thing to select oculars with the urban environment in mind. Wherever you observe from, though, you can't go wrong choosing the best eyepieces you can afford. No matter how many times you change telescopes over the years, your eyepieces will still be useful. Unless you are interested in the most expensive ultrawide field eyepieces, choosing "best" rather than "cheapest" does not cost much more. For example, the excellent TeleVue Plossl eyepieces currently cost a little over twice what you'll pay for the average Chinese Plossl. Sometimes imported eyepieces can be good optically, but rarely as good as a TeleVue or other top brand. You can also be assured that name-brand oculars are better mechanically as well as optically, and are well suited for the long-haul. "Twice as much" for a considerable increase in real quality doesn't seem bad when you consider the fact that you may be living with an eyepiece for several decades.

Eyepiece Characteristics

Eyepieces, like telescopes, can be described with just a few numbers. Most important is the eyepiece focal length. Eyepieces (also often called "oculars") are commonly found in focal lengths of 40-mm down to about 6-mm, with 50–55-mm and 5–3-mm models also readily available, if not as popular. An eyepiece's focal length lets you know its magnification potential. The shorter the focal length, the higher the magnification in a given scope. To find an eyepiece's "power," as explained in the telescope section of this book, divide eyepiece focal length into telescope focal length. A 10-mm eyepiece gives you 200× in a 2000-mm focal length scope and 100× in a 1000-mm focal length instrument.

"True field" is the amount of sky you can see with your telescope and eyepiece combination. To find true field, you divide the magnification given by an eyepiece in a particular scope into its "apparent field" value (which should be listed in the manufacturer's specifications for that eyepiece). An eyepiece with a magnification of 200× and an apparent field of 85° yields a true field of 0.42° (85/200 = 0.42), a little less than half a degree of sky.

The aforementioned apparent field figure describes the expanse of field visible to the eye when looking into an eyepiece. Don't confuse this with true field. While the expanse of space visible to your eye may extend across 50° of your field of vision, for example, this may only encompass half a degree of *real* distance in the sky. Eyepieces with large apparent fields are more comfortable to use and more impressive (TeleVue's Al Nagler calls this the "spacewalk experience"). Using an eyepiece with a large apparent field compared to an eyepiece with a small apparent field is like watching a program on a 70-inch projection television rather than a 12-inch portable.

Although there are dozens and dozens of eyepiece brands available to amateur astronomers in the U.S. and Europe, the choice of an ocular is actually fairly simple. There are only four basic families of designs that are widely available and popular at this time, and many of the multitudinous brands advertised in the astronomy magazines are actually, like many of today's scopes, rebadged Chinese eyepieces from the same factories.

Eyepiece Designs

Many eyepiece types have come and gone over the 75 years since amateur astronomy became a popular pursuit, but all that experimenting has finally boiled down to the four basic designs/types shown in Figure 3.1. The simplest, the three element Kellner, is popular with amateurs for one reason: it's cheap. These eyepieces work fairly well in longer focal lengths (20-mm and longer), but are less usable in short focal lengths due to small eye relief—you have to place your eye very close to the lens to see the entire field, a problem for eyeglass wearers. The edge of the field is not very sharp in a Kellner, either, especially with smaller focal ratio scopes. Kellners can do well in large focal ratio instruments like SCTs, however (large focal ratio telescopes are always more forgiving of eyepiece deficiencies than small focal ratio ones). Kellners seem to be on their way out lately. With the advent of dirt-cheap Plossls from the Far East, telescope manufacturers can afford to include "better" eyepieces with their new scopes, and there's little reason for an amateur to buy a Kellner when a Plossl is only a few dollars or pounds more expensive.

Like the Kellner, the *Orthoscopic* design has been around for well over a century. These eyepieces are good performers in long and short eyepiece focal lengths, and

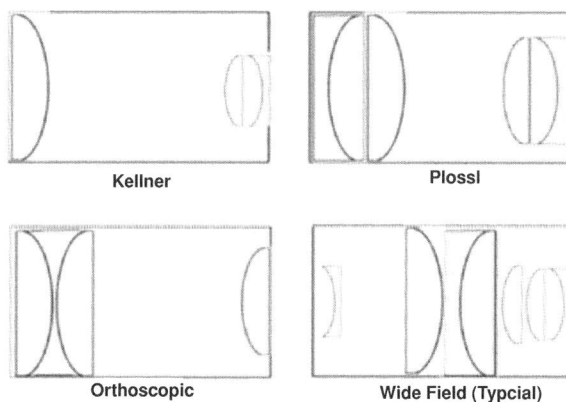

Figure 3.1. Popular eyepieces designs.

have sharp, flat fields. Their main drawback for the deep sky observer is their small apparent fields of view. It's like looking through a keyhole when you are restricted to 40–45° common in Orthoscopics. This doesn't bother planetary observers, of course. In fact, the Orthoscopic remains the eyepiece of choice for observers who don't need wide apparent fields. Its innate sharpness and the fact that it places only a few lens elements in the light path (four), means that the "Ortho" will likely remain a favorite for many years to come.

The Plossl (also known as the "symmetrical") is *the* eyepiece these days. It is without doubt the most popular ocular design with amateurs. Why? It's a good all-round performer, presenting sharp, flat fields across the entire range of focal lengths. It is also, as mentioned above, beloved of far-eastern eyepiece makers and is thus available for very modest prices. It performs equally well on the Solar System and on the deep sky, and, while not a wide field design, delivers a comfortable 50–55° apparent field. The perfect eyepiece? No. Its main drawback is short eye relief in short (less than 10-mm) focal lengths. This four-element eyepiece is now the obvious choice for observers, especially those on a budget.

The *premium wide field* eyepieces, represented by the TeleVue Naglers and Panoptics, the Meade Series 5000 Superwide and Ultrawide eyepieces, and Pentax's XW eyepieces, are a very different experience. Their incredibly expansive apparent fields—ranging from about 70° to 85° or more depending on the exact design of the eyepiece—make for an amazing experience. Looking through one of these expensive eyepieces (expect to pay at *least* 250 US$ or 200 UKP for the *less* expensive models) spoils you. It's hard to go back to peering through the "peephole" of a Kellner or Plossl after the picture window of a Nagler or Ultrawide.

If these eyepieces have a failing other than high prices, it's that they tend to be short on eye relief. TeleVue is working to improve this characteristic in its current designs, and this is a good thing, since it is very frustrating for an eyeglass wearer to own an eyepiece with a huge apparent field, but not be able to see all of it at once. These optically sophisticated eyepieces also put a lot of glass between you and your deep sky object. Some of the Naglers are made up of as many as seven lens elements, and images in a simple Kellner or Plossl may be noticeably brighter despite the advanced coatings used by wide-field eyepiece makers. In general, this brightness penalty is minor and is outweighed by these eyepieces' other advantages.

Eyepiece Considerations for City Lights Astronomers

Bathed in sodium streetlights, our need for long focal length eyepieces is limited. In general, you'll want to stick to the medium focal lengths, usually, with a 25-mm rather than a 35-mm being your low power eyepiece. In fact, I find that my most used eyepiece focal length in light-polluted areas is 12-mm, even with my long focal length SCTs. I'll use a 22-mm for my finding eyepiece, since it offers a fairly wide field that's not saddled with a background that's so bright as to obscure objects, and then switch over to the 12-mm for serious viewing. A good beginning set of eyepieces for the city observer might be a 25–20-mm "finding eyepiece," a 15–12-mm "workhorse," and a 7–6-mm "high power" for the small galaxies and planetary nebulae.

So, don't buy a 35-mm eyepiece. And get a 12-mm instead of a 20-mm as a medium power ocular, right? Maybe not. There is one instance where you might want longer focal length eyepieces. If you need good eye relief, rather than paying a premium for a shorter focal length eyepiece designed with this characteristic in mind (like the Tele-Vue Radians and the Vixen Lanthanums), just "barlow" a longer focal length eyepiece. Barlow lenses, shown in Plate 11, in their simplest form are single element negative lenses. Place one ahead of your eyepiece and it increases your magnification (typically by two or three times depending on the barlow's design). Long focal length eyepieces typically have better eye relief than shorter ones, so you can gain a comfort advantage by using an inexpensive 25-mm eyepiece barlowed to 12.5-mm (2× its original magnification) rather than an expensive 12.5-mm LER "Long Eye Relief" model.

Some novice amateurs are skeptical of barlows. How can you double your set of eyepieces with an inexpensive barlow? There *must* be a catch. Actually, no, not this time. A good barlow—and all of those I've tried from major manufacturers are outstandingly good these days—doesn't have any drawbacks. A modern two- or three-lens element barlow doesn't hurt your image quality. In fact, if you use simple eyepiece designs with a small focal ratio telescope, a barlow can actually *improve* image sharpness at the field edge. A barlow is a must-have accessory.

Premium Wide Fields or Not?

Should you pay the big money for a Meade, Pentax, or TeleVue wide-field eyepiece? If you can afford it, yes, in my opinion. These eyepieces are painfully expensive for most of us, but, as mentioned earlier, you will be able to use one for the rest of your observing life. In addition to their comfortable and impressively wide apparent and, therefore, true fields of view, they offer a real advantage in the city: their wide fields make it very easy to star-hop to objects in the main eyepiece. In some areas of the sky even my 50-mm finder doesn't pull enough "signposts" out of the glow to allow me to easily find deep sky targets. The area of the Virgo-Coma galaxy cluster is an example. What do I do? I pop in my 12-mm Nagler or 22-mm Panoptic (depending on the focal length of the scope I'm using). These eyepieces provide enough magnification to darken the field dramatically, but display a wide enough swath of true field allow me to jump from star to star with the aid of detailed charts on my way to galaxy after galaxy. I'd rather have one or two premium eyepieces than a whole boxful of cheap Plossls—especially in the city.

Roadmaps to the Sky: Atlases and Charts

In city or country, you need detailed sky maps to find anything. Forget the simple all-sky charts found monthly in the astronomy magazines. They don't show enough stars to allow you to locate anything beyond the brightest objects. An adequately detailed atlas is actually often far more important for the urban astronomer than for the country observer. If you've got dark skies, you can often just get in the neighborhood of an object. The target will be visible in the finder, making it simple to place it in the field of the main scope's eyepiece. In the city, all but the brightest DSOs will be invisible in finders, even large ones, so you'll need to be "on the money" every time. Only detailed charts, printed or computer-generated, can help you do that.

Print Atlases

There are quite a few print star atlases available to the amateur. Although computer charting is popular among astronomers now, traditional paper atlases don't seem ready to disappear. There are actually more of them on sale than ever. As far as I'm concerned, the perfect atlas for the urban observer has yet to appear. My dream atlas would be composed of several series of charts, with each series covering the entire sky. There'd be one group with a limiting magnitude of 3, one for 4, and maybe one for 5. Each of these series would be linked to a detailed group of charts showing stars down to magnitude 8 or 9 and thousands of DSOs (remember, the higher the magnitude number, the dimmer the star). An Australian atlas, *The Herald Bobroff AstroAtlas*, which has just come back into print after being gone for a while, almost fulfills these requirements, but its series of charts are not quite what I want either in magnitude limits or completeness. Since my dreamed-of *Urban Star Atlas* has yet to be produced, we'll have to settle for the next best thing.

Take a look at *Herald Bobroff* if you can find a copy—it's an interesting atlas—but I still consider "the next best thing" to be *Sky Atlas 2000* by Wil Tirion and Roger Sinnott (Plate 12). This large format set of charts covers the sky down to magnitude 8.5 for stars and features many thousands of DSOs. It's about as perfect as it gets for the city. Nearly every star you can see in light-polluted areas through a 50-mm finder is plotted, and the selection of DSOs is quite appropriate for what is doable in the urban environment. *SA 2000* comes in several editions: an unbound Desk version with white sky/black stars, an unbound Field version with black sky/white stars, and a larger format Deluxe edition that is spiral bound and features color and a white sky. *SA 2000* is the basic atlas for the urban observer in my opinion.

Other options? Many "brighter" atlases that only show stars down to magnitude 6, are available. Leave these alone. You need a chart that includes all the stars you can see in your finder. There are more detailed atlases, too. *Uranometria 2000* by Wil Tirion, Barry Rappaport and George Lovi, reaches magnitude 9.5. Roger Sinnott and Michael Perryman's *The Millennium Star Atlas* goes all the way to magnitude 11. Both of these "super atlases" are incredibly beautiful works, even if they might seem to be overkill for the urban observer. Why pay the high prices these books command when you won't be able to see the thousands of PGC and UGC galaxies they plot? That's what I used to think, anyway. Then I discovered how useful the super atlases could be for star-hopping through the main eyepiece. When an atlas shows most of the stars visible in a medium power eyepiece in your main scope, it's pretty easy to move from object to object without ever using a finder. You'll still want good, old *SA 2000* as your main tool, though.

Computer Planetariums and Atlases

Despite the continuing presence of print atlases, there is no doubt that computer star mapping programs are becoming more popular with astronomers every year, and are taking over many of the duties traditionally reserved for books. Some amateur astronomers are reluctant to make the switch since they don't have a laptop computer to use at the telescope. But owning a laptop is not necessary to gain the benefits of computerized charting, since all current programs will print hard-copy charts that are very close to the quality of the best printed atlases.

Computer charting programs are particularly well suited to urban observing since you can easily tailor their displays to urban conditions. Your limiting magnitude is 4? It's simple to tell a computer program to plot *only* stars of magnitude 4 or brighter. Want to display all the stars visible in your finder? Again, easy. It's also a snap to create highly detailed and correctly oriented eyepiece field-sized maps to allow you to star hop with your main scope. As far as resources go, current computer atlases contain far more detail than even the best printed atlases. The program *Skymap Pro*, for example, contains stars down to magnitude 15, while *Millennium* leaves off at 11. Deep sky objects? *Millennium* has a respectable 10,000, but Skymap easily overwhelms it with 200,000.

Introductory Planetariums

The astronomy programs available now have divided themselves into two broad general classes: planetariums and planners. Planetariums, especially inexpensive entry-level ones, are what most of us gravitate toward when choosing a first astronomy program. Their operation is easy to understand; the program creates a representation of the night sky on your monitor screen. Buttons or menus allow you to change the time of day/date, direction of view, zoom in and out, and perform a few other functions. Entry level programs typically plot stars down to magnitude 9 or 10 and display all the Messier DSOs and a selection of the brighter NGCs. This level of detail is adequate for a beginning observer, but you'll soon be left wanting more. Even in light pollution you're likely to soon begin detecting little galaxies, for example, beyond the range of a beginning planetarium. Also, introductory level programs tend to concentrate more on presenting a "pretty" depiction of the sky than on creating printed charts that will be usable and legible under a dim read light. Still, for the money (about 30 pounds or dollars on average), one of these will get you started using the computer with the scope. Current examples of beginning planetarium programs are *Starry Night Backyard* and *The Sky Student Edition*.

Advanced Planetariums: Deep Sky Software

Moving beyond the most simple and inexpensive planetariums, we come to the programs that have made astronomy software so popular with deep sky observers. These advanced planetariums (often referred to as deep sky programs) do the same basic job as the simple ones—they build a night sky on your computer screen—but in far more detail. The average advanced program will plot *millions* of stars—usually the entire Hubble Guide Star Catalog down to magnitude 15 and dimmer. The latest releases go far beyond the Messier for DSOs, many featuring up to one *million* nebulae, clusters, and galaxies.

Urban observers won't be able to see even a tiny fraction of this million fuzzies, but the advanced programs have other advantages in addition to object totals. They include numerous tools for the observer that simpler programs lack. You can, for example, rotate and flip onscreen or printed charts to *exactly* match the view your scope produces at the eyepiece. These programs can even draw circles on the screen (and on printouts) representing the sizes of the fields of the exact eyepieces used in *your* scope. This makes it easy to produce eyepiece-field size finder charts. It is now very common for this level of software to also allow you to control a go-to scope from the computer. Click on an object onscreen and the telescope moves to it. These features come at a slightly higher price than the introductory programs, with advanced planetariums selling for 100–150 dollars or pounds.

Specific programs that have proven popular with today's serious amateurs in and out of the city are *Skymap Pro*, *Megastar*, *Guide*, *The Sky 6*, and *Cartes du Ciel*. *Cartes* is particularly noteworthy in that it is *freeware*. While it is very capable, offering the same kinds of features as the other programs in this class, the author, Patrick Chevalley, has chosen to distribute it for free over the Internet (see Appendix 1 for the *Cartes du Ciel* URL). Why give away such a wonderful program? Patrick is committed to amateur astronomy, and says he'd rather see amateurs spend their money on eyepieces than computer software.

Planners

The most useful type of astronomy program for the urban observer isn't really a planetarium at all. It's a database. Yes, that sounds dry and uninteresting, but you'd be surprised how much computerized astronomical databases—the planners—can do for you, and how much fun they can make doing it. Like the advanced planetariums, they usually contain the Hubble Guide Star Catalog and up to a million DSOs. The difference between them and the planetariums is that they allow you to *manage* all this data in order to help you plan observing sessions (see Plate 13 for a typical planner display).

Only want to observe galaxies brighter than magnitude 10 that are on the meridian at 8 pm for your location? These programs will easily produce a sorted observing list containing *only* the objects that meet that criterion. The lists created by planners will contain all the details you need to find and study objects—coordinates in Right Ascension and declination, descriptions, magnitudes, and much more. You don't have to give up charts, either. Most planners will produce very usable charts at the same level of detail as the advanced planetariums. The charting "modules" of the planners *usually* don't have quite as many features—especially navigation buttons—as full-blown planetariums, but they are still quite capable. Some of the most recent planning programs incorporate advanced features like go-to telescope control and image libraries. Planners are the future of advanced astronomy software in the city or the country.

Two programs currently rule the roost when it comes to planning/database software for the PC: *Skytools 2* and *Deepsky Astronomy Software* (DAS). Both are similar in the number of stars and objects they contain (millions) and features (many), but each has its own strengths. *Skytools 2* provides a very user-friendly interface and has a "photo-realistic" charting mode whose displays can rival those of hyper-realistic planetarium programs like Starry Night. *Deepsky's* strength is in pictures. The program includes a supplemental DVD disk that is filled with over 400,000 images of DSOs. That means most objects you can click on in the database or on a chart will have a photo associated with them—very helpful in identifying faint fuzzies. The program is also available in a CD-only edition, which includes a CD with a smaller but still useful number of objects—11,000. Don't have or want a PC? *AstroPlanner* is a planning program for the Apple Macintosh that is inexpensive and quite capable. Its charting capabilities are somewhat less advanced than the PC software, but usable, and it offers some interesting features like helping go-to scope users in choosing "good" alignment stars.

Miscellaneous Accessories

Scope, eyepieces, charts? Anything else? Yes, you'll need a few other items to make your observing life fun and comfortable.

Red Flashlights (Torches)

Sure, the sky glow in the city is almost bright enough to allow you to read your charts without further assistance. Almost. You'll need a decent red light to enable you to read your star maps without ruining what night vision you're able to attain in your bright environment. Some users make do with a standard white light covered with red paper or plastic. That doesn't work very well. Usually, the light produced is not very red and is much too bright. My favorite type of astronomy light is one that uses red LEDs. These little devices are very inexpensive from astronomy dealers, produce pure red light, and have controls that allow you to adjust the brightness. Most use two LEDs, and can produce enough light for almost any purpose. Some even allow you to switch in a pair of blue LEDs when needed—as when walking back to the house at 3 a.m.

Tables

You'll need somewhere to put all these accessories. I favor a simple lightweight folding camp table. These can be found in sporting goods stores. If you don't bring too many items outside with you, a card table can also work well.

Observing Chairs

Might as well be comfortable. If you own a refractor or an SCT, you may even be able to do all your observing while seated. Being comfortable, you'll find, can really increase your stamina and allow you to do very productive observing far into the night. Quite a few custom made "observing chairs" are available from astro-dealers, but I find what works best for me is a simple and inexpensive drummer's throne (stool) from the local music store.

You can—and probably will—keep on buying, but what we've accumulated so far will serve you well when we begin our walking tour of the cosmos. But before getting started let's talk about *how* to observe in the city.

CHAPTER FOUR

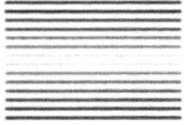

Urban Observing Techniques and Projects

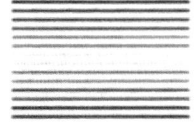

All amateur astronomers, as they grow and progress in this wonderful hobby, develop a "bag of tricks," techniques that allow them to squeeze every last detail and every last photon out of their telescopes. In the city, it is especially important to take full advantage of all the light our scopes can gather. In large part, taking advantage of everything your scope can do for you is a matter of experience. Observe more and you'll see more. But there are some specific techniques you can learn that will improve your city observing results.

Averted Vision

Most deep sky observers are well aware of a special characteristic of the human eye: it is more sensitive on the periphery of its visual field than it is at the center. The retinas of your eyes are composed of two different types of light receptors, rods, and cones. Rods, which are found in greater numbers away from the center of the retina, are considerably more sensitive than the cones. The cones handle color vision, while the rods are essentially color blind, explaining why it is so hard to see color in dim astronomical objects. If an object is so dim that you need your rods to see it, you just won't detect color. The human eye and the image processing done by the brain is a very complex subject, and this description of the eye's light-sensing characteristics is grossly oversimplified, but the fact is, you can see dimmer details of an object (or even make an invisible object appear) by *looking away from it* rather than *right at it*. This is averted vision.

The funny thing is that amateur astronomers who know this technique and use it regularly while observing from dark country skies seem to forget to use it in the city. While the improvement in your ability to see dim objects and details using averted vision may not be quite as dramatic with bright-background city viewing, the gain is still quite real and can result in your seeing a dim galaxy with ease instead of adding it to a sadly long "not seen" list.

Jiggling the Telescope

There's another characteristic of the human eye you can take advantage of in your quest to see faint objects. The human eye can distinguish moving objects more easily than stationary ones, an adaptation that may have allowed our distant ancestors to quickly detect moving predators out on the veldt. If you can't make out an object with averted vision, try gently tapping on the tube. You may find you can see your dim target as long as the telescope is vibrating.

Dark Adaptation

Have you ever gone from a bright sunny afternoon into a dark motion picture theatre? Once inside, you are not surprised to find that you're practically blind for a while. After about 5 minutes you'll find your sight returning, with colors becoming more visible as your cones adapt to the darkness. Over the course of approximately 30 minutes, your rods attain their full sensitivity. Slight improvements in your dim-light vision may continue for a considerable length of time after that. The adjustment of the rods and cones to dim light is caused by the production of a chemical called *rhodopsin* or "visual purple." The amount of this substance in your rods and cones determines how sensitive they are to light. Exposure to bright light bleaches rhodopsin out of the eye and causes it to become temporarily less sensitive to light.

Dark adaptation is actually a two-part process. In addition to time in the dark needed to allow your eyes to recharge their rhodopsin, your eyes' irises also have to have time to dilate. Your irises are small diaphragms that can open or constrict to let in more or less light, ensuring that your retinas receive a sharp and properly "exposed" picture. The iris of your eye works exactly like the f/stop diaphragm in a camera, which was modeled on the working of the human eye. A normal human iris will open from 6 to 8-mm depending mainly on the age of the observer (older peoples' irises open less wide than those of youngsters). For your eyes to attain their best dark adaptation, they need to be in total darkness for at least 30 min. Unfortunately, that's tremendously difficult to do in the city.

As was pointed out earlier, there's nothing you can do about city sky glow. It's there, and that alone will prevent you from reaching full and complete dark adaptation. But if sky glow were all you had to worry about, your eyes could actually attain a good degree of dark adaptation in most locations. What prevents dark adaptation in the city is ambient light. A neighbor's nearby security light or even the light shining from a curtained window will radically reduce your eyes' final dark adaptation. They'll only become sensitive to dim light *to a degree*, and if you accidentally look straight into a streetlight or security light, your dark adaptation will have to start over from square one. Fortunately, ambient light can be dealt with.

One approach, as discussed in the site selection section, is to choose an observing location with minimal ambient light—no bright lights shining directly on you and, if possible, nothing other than the general sky glow illuminating your and your telescope. Naturally, this is often impossible in the urban environment, where dozens of ambient light sources are present. The next best approach is *blocking*.

The idea is to shade your telescope from all ambient light. Put up a shield of some kind. This shield can be as simple as an opaque sheet of vinyl or cloth hanging on a line strung between two posts. Almost anything can be used as a light shield, from a sheet of plywood propped on a step-ladder to a large piece of cardboard hung from a tree limb. Some manufacturers have gone so far as to produce portable temporary observatory domes made from tent canvas. These can be extremely effective, but assembly is a complicated annoyance if you have to put them together for every viewing session. Most, however, are durable enough that they can be left erected for weeks during favorable observing weather if local codes/covenants and your spouse allow.

I didn't want to spend the money for one of the tent-type observatory domes, but recently got tired of trying to prop pieces of cardboard or plywood against a step-ladder and trees. While my shields worked well, they were difficult to move around, and a sudden gust of wind would usually send the "wall" of my "observatory" crashing into my scope or my head. I came up with a design for an easy to use, movable light-shield one night as I was attending a play at the theatre. Why not use stage flats? Flats, painted stage scenery, are made of canvas stretched on a frame of 1 × 4 lumber, and are held in place by a brace and a sandbag. This sounded like something that could be adapted into an easy-to-move light-shield. Stagehands move scenery all the time, and in the smaller sizes I'd use (stage flats can be as much as 12 feet high), portability wouldn't be an issue. As a test, I constructed a 4 foot wide by 6 foot high frame of 1 × 4 lumber connected at each corner with a triangular piece of $1/4$ inch plywood held in place by wood glue and screws. A triangular brace, also made of 1 × 4 lumber, was attached to one vertical 1 × 4, and was held in place with a concrete block or bricks (though sandbags as used in the theatre will work if you want to go to the trouble to make them). A piece of inexpensive muslin from a fabric store was stretched over this frame and stapled in place. My "flat" was then painted with enough coats of black latex paint to make it impervious to light.

In use, I position my flats/light shields (Plate 14) as needed to keep ambient light off me and my telescope. If you have severe ambient light problems, you may want to construct three or four or more of these shields. They can be linked together with door hinges (perhaps with their pins replaced by removable rods or nails for easy disassembly) or eye and hook latches to provide a sturdy shield. By using inexpensive pine lumber I kept my cost to a minimum, and, considering the small expenditure, the improvement in my observing conditions has been dramatic. In addition to blocking stray light, my well-braced stage-flat "observatory" also helps protect the scope from wind, ensuring steady views on those good winter nights after a front has passed through.

If you don't like my stage flat observatory idea and want something more elegant, you might make shields with frames constructed from PVC pipe. This can be arranged so that the pipes screw together, meaning the light shield can be disassembled for storage. Instead of muslin, a lightweight, opaque vinyl tarp, possibly attached to the pipe frame with elastic cords, might be nice. But my humble stage flats work incredibly well for me.

Speaking of observatories, that's not a bad idea. Dark sky observers would probably find the idea of a permanent urban observatory hilarious, but an observatory is perhaps even more useful in the city than in the country. In the city, a real observatory provides the same great luxury it does in the country: a permanent place where you can leave the scope set up so you can observe happily at a moment's notice. Not having to haul the scope and accessories in and out means you will observe a *lot* more, especially on weeknights. In addition, a properly designed observatory can end your problems with ambient light forever, effectively blocking everything but the night sky.

An urban observatory can be built exactly the same way as a country one, with roll-off roof models like the one shown in Plate 15 being easier to make and more popular than traditional domed installations. The only city-peculiar considerations are a possible need for higher security, which can be easily provided by adequate doors and locks, and a need to deal with building codes, which are not often a hindrance out in the country. If security is a consideration for you, I'd suggest that you might want to forego a traditional dome and do a roll-off roof observatory. A dome is beautiful, but calls attention to itself, and may be an irresistible magnet for vandals and thieves.

It is beyond the scope of this book to show you how to build an observatory, but the process of putting up a secure, attractive, and useful building is not difficult, especially if you have friends experienced in building or remodeling. Even if you don't have access to someone with that kind of talent and don't have it yourself, you may not be destined to do without an observatory. Your local library will have books in its "home improvement" section that provide step-by-step instructions on constructing small outbuildings and sheds. Combine these plans with details of amateur observatories found in amateur astronomy books and magazines and your long-dreamed-of-facility can become a reality. If this seems too daunting, it is also possible to modify the prefabricated outbuildings that are available from many sources. Your observatory does not have to be fancy; the only requirements are that it provides a secure, dry home for your scope, shield you from ambient light, and allow you to see as much of the night sky as possible.

Observatories are only an option if you live in an unattached dwelling with a backyard. If you must travel to public or semi-public locations to do your urban sky-watching, even light shields may not be practical unless you can some up with some kind of collapsible design like the PVC pipe frame idea. But there are still ways to defeat ambient light, especially if you don't mind looking a little ridiculous. How can you shield at least one eye from ambient light? With an eye-patch. Keep an eye-patch on your dominant/observing eye while you are not looking through the scope. When you're ready to view, flip the eye-patch up. Eye-patches are available from the neighborhood pharmacy/druggist.

But how do you keep the light out of your "good" eye *while* you are observing? Even with your eye pressed to the eyepiece, the glare from bright ambient lights still intrudes. Use a hood. Find a piece of black, opaque cloth—I like rip-stop nylon—and drape it over your head while you are at the eyepiece. Don't flip up your eyepatch until the hood is over your head. You will be surprised, amazed even, at how much these two simple items can do to improve your views. I guarantee that if you use an eye-patch and a hood in ambient-light-plagued areas, you will see *much* more than you otherwise would.

What if you don't like wearing an eye-patch? It's not very comfortable. And you may feel that it makes you look slightly odd (who cares?). A more comfortable alternative is a pair of red goggles (Plate 16). Obtainable from many astronomy dealers, these "light pollution goggles" are really safety glasses that have been equipped with red lenses. Use these when you're not looking through the scope, and both eyes will be protected from dark-adaptation-destroying ambient light. When you want to look through the scope, you should still, naturally, use a hood.

Red goggles, which can be easily made if you can't find them for sale, work for the same reason that red flashlights do. The rods in our eyes, the dim light receptors, are least sensitive to red light. In addition to keeping both eyes dark adapted, goggles have another advantage over an eye-patch—they preserve your 3D vision—you lose that with one eye covered. This is an important safety consideration when you have to move around. Without 3D vision, you're likely to trip over something or bump into your scope hard enough to ruin your alignment.

A possible disadvantage of these goggles is that high levels of even red light will adversely affect your dark adaptation. Accidentally looking into a streetlight with your goggles on will mean you have to start dark adaptation all over. I have also found them prone to fogging in my humid environment, despite the fact that they are provided with ventilation holes. Finally, you won't look much less ridiculous with red safety goggles on in the middle of the night than you would with a simple eyepatch (if such things matter to you).

Optimizing a Telescope for Urban Use

As delivered, your telescope may not be optimum for use in the city. The problem is that unless you take steps to prevent it, your scope will admit stray light into its optical system. No matter how well you try to shield your telescope from ambient light sources, there's still a lot of unwanted glow in the urban environment. Even if the scope is shaded from nearby streetlights, the glow from the sky may be leaking into your tube in unexpected ways. This stray light can have a very detrimental effect on your deep sky views. In my experience, eliminating it can have almost as much effect as allowing your eyes to dark adapt with hoods and eye-patches.

Dew Shields

Unless you observe in a big open space like a park or parking lot, dew is not usually a problem in the city. Nearby houses, buildings and trees tend to block your scope from the heat-sucking sky and prevent the resultant formation of dew on your lenses or mirrors. A dew shield is a must for the urban observer nevertheless. In addition to protecting your optics from dew, a dew shield can prevent oblique ambient light from entering the corrector or objective end of your refractor or SCT and ruining your images.

If you've got a refractor, it likely came already equipped with a dew shield, but that doesn't mean it's perfect. Many of those on current import refractors suffer from severe problems. Most serious is that they are too short. To be effective, a dew shield should extend in front of your objective for at least 1.5 times the lens diameter. A 6-inch refractor, for example, needs a 9-inch long dew shield. If your dew shield does not measure up, replacements are available from various sources, or you can make your own very easily. Posterboard or cardboard that's been painted flat black and sealed makes a fine dew shield. One made out of the black foam padding material sold to go under campers' sleeping bags works even better.

Another fault of the stock dew shields on refractors is their color. Look at the inside of your dew shield. Is the finish even the slightest bit shiny? If so, it is probably reflecting a lot of unwanted glare into your optical system. One solution is to paint it a very flat black. Black paint sold for use on bar-b-que grills is commonly available—at least in the U.S.—and is just right for this purpose. Even better than a new paintjob, however, is lining the dew shield with a black fabric like velvet. This can be affixed to the interior of the dew shield with contact cement, or a visit to a fabric or craft store may turn up some black self-adhesive velvet-like fabric or contact paper. Using velvet to line your dew shield will have the added benefit of absorbing some moisture if your site is prone to heavy dew.

If you're the proud new owner of a Schmidt Cassegrain telescope (SCT), your instrument came without a dew shield of any kind. There are many dew shields offered for sale to SCT users. Most have one thing in common: they are too expensive. You can easily spend 150 US$ or more on a fancy aluminum shield painted to match your telescope's OTA. If you want to spend that much money, go ahead, you won't hurt my feelings. Keep in mind, however, that one of these expensive models won't work any better than a piece of posterboard or plastic sheeting. Me? I'd rather spend my money on a new eyepiece. If you don't want to make your own dew shield, astronomy dealers also offer "flexi-shields." These are flat pieces of plastic that can be wrapped around the end of your tube and secured with Velcro to form a dew shield. These typically cost about a third as much as one of the aluminum models.

If you own a Newtonian, you don't need a dew shield, right? The mirror is at the bottom of the tube, making your entire OTA one giant dew shield. Wrong. Your telescope has its secondary mirror at the forward end of the tube where it is nearly as susceptible to stray light and dew as the lenses of refractors or SCTs. This wouldn't be a problem if Newtonians had a decent length of tube ahead of the focuser, but today's short focal length reflectors, especially the imported models, tend to mount the secondary and focuser as close to the end of the tube as possible. Do yourself a favor and extend the front of your tube at least 12 inches. This will keep stray light out and keep dew off your secondary. Use the same approach to making a tube extension as for making a dew shield—cardboard, posterboard, or plastic sheet painted flat black and sealed if necessary.

The front end of the tube is not the only place where unwanted light can enter the tube of your Newtonian. The back end, the mirror cell end, also provides a path for glare. If your OTA features an open style mirror cell—if you can look down the front of the tube and see the ground through the back of the scope—unwanted light is intruding into your scope. The sky glow reflected off the ground and back into this rear opening in the tube is enough to *substantially* degrade your images.

The solution is to construct a baffle for the back of your tube. This can be as fancy or simple as you care to make it. All you need is a round disk of some material—paper, plastic, wood, metal—with a hole cut in it slightly smaller than the diameter of your mirror. As shown in Plate 17, this baffle blocks the opening around your mirror's edge, but leaves the mirror cell open to the air, allowing it to "breathe" so that it will acclimate to external temperatures more quickly than it would if you sealed off the rear end of the tube completely. Many of the current Chinese Newtonians have a thin metal plate sealing the mirror end of the tube. This does a good job of blocking light from the ground, but should be removed and replaced with a baffle in the interests of rapid cooldown.

One final thing to look at on Newtonians is the paint job on the tube's interior. Is it really flat black? Can you see *any* reflections? Is it actually more of a dark gray than a black? If so, you'd be wise to repaint the interior of the tube if possible. Select a flat black paint and apply enough coats to completely squelch the shiny reflections (remove your optics beforehand, of course). To really suppress internal glare, mix a little sawdust in with your paint to give the finish a rough texture and further inhibit reflectivity. If painting the inside of your scope's tube does not appeal to you, there is an easier solution that is almost as effective. All you have to do is glue a piece of black velvet to the interior of your tube exactly opposite your focuser. This will at least prevent the area near the front of your tube from reflecting ambient light or sky glow directly into your eyepiece. I applied all these fixes to my 1960s vintage 4.25" Newtonian and was amazed at its improved images. No, it didn't become the equivalent of an 8-inch scope, but the sky background was noticeably darker, and deep sky objects (DSOs) stood out much better.

Using Eyepieces in the City

If you've read this far, you've gathered that I recommend using higher magnifications in the city than in the country. But what does that mean? Exactly how much magnification should you use for finding and viewing objects? When it comes to finding objects, much will depend on your scope, your site, and your finding skills. But what you want to do is find an eyepiece/scope combination that provides a reasonably dark but reasonably wide field. This is obviously a compromise. Even if you are an experienced "star hopper" able to locate any object in seconds under dark skies, the lack of guide stars in bright city skies will mean that you will not be as accurate—you'll have to hunt around. Obviously, a low power eyepiece, a 35-mm or 40-mm wide field design would be a big help, shortening your search. Unfortunately, at low powers, you may never find your objects. The sky background is bright enough to obliterate a dim galaxy, for example.

Settling on a finding eyepiece is a matter of experimentation. If you've got a good collection of oculars, train your scope on a fairly challenging medium-sized DSO (M51, the Whirlpool Galaxy, is a good one for most urban observers using 8-inch and larger telescopes). Keep lowering your magnification by going to successively longer focal length eyepieces until you find one that allows you to *barely* discern that the galaxy is in your field. Don't worry about details, you just want to be able to see that the galaxy is there. The ocular that provides this magnification will be your finding eyepiece. If you are hunting easier objects—open star clusters rather than galaxies and nebulae—you can lower your power to make things easier.

Even if you have a go-to scope, you'll want to use as wide a finding eyepiece as practical for your scope. Current go-to-equipped instruments are surprisingly accurate when carefully aligned, but, as was mentioned earlier, you cannot always rely on them to always put an object in the field of a medium power eyepiece. The widest field eyepiece you can use effectively in the city will make object location with a go-to scope much quicker if your target doesn't appear in the field at the end of the slew. You won't have to search around using your hand controller to slew the scope at low speed.

Once you've got a target object in view, how much magnification should you use for observing in light-polluted areas? Like the choice of the finding eyepiece, this is dependent on several factors: scope, sky, object type, and observer preference. In general, try to use as high a magnification as possible that still frames the object well— lets you see as much of it at one time as you desire—and provides a comfortable view.

Once again I'll emphasize: deep sky observers in the city *or* in the country tend to use less magnification than is optimum. The human eye is capable of making out details more easily in a large image than a small one. You don't, of course, want to use so much magnification that the object becomes dim to invisibility. You want to find a power that darkens the sky background enough to provide good contrast while keeping the object in view as bright as is possible. Experimentation will show you what's right for you and your scope, but I find myself most often using about 100–150× as an everyday viewing magnification, and will often go considerably higher—to 200× and above—for small galaxies and planetary nebulae. This is with my 8-inch and larger scopes. For small apertures, you may have to scale back these magnification values somewhat.

Binoviewers

This subject could easily have gone into the Accessories chapter, but binocular viewers for telescopes are such a specialized piece of equipment that I thought they'd be better off here. As shown in Plate 18, a binoviewer is a device that is inserted into your star diagonal (refractor or CAT scope) or focuser (Newtonian), and which uses a system of prisms to divert the light from your telescope's optical system into two separate eyepieces. What good is this? All day long you use *two eyes*, but when night comes and you drag out the telescope, you use *one eye* to view the heavens. Viewing with one eye is unnatural and uncomfortable whether you are in the dark country or bright city. Some people can train themselves to leave their unused eye open while viewing through a telescope's eyepiece, and that helps some, but is very hard for most observers to do in the less than dark environment of the city. Most amateur just squint painfully. Binoviewers are offered as a remedy for this "wasted eye syndrome," and makers and fans of these accessories swear that you can see more with two eyes than you can with one. I'm always looking for ways to squeeze every last photon out of our city skies, so I decided to see what a binoviewer could do for the urban observer.

The U.S. company Denkmeier Optical was kind enough to loan me one of their binoviewers for evaluation purposes, and I found it to be a very well constructed, high quality piece of gear. I located a couple of identical 25-mm eyepieces to use in the Denkmeier and hit the backyard for an evening of comparative viewing. What I found was that the binoviewer *did indeed* reveal more details in any DSO I put into the field of view. The image in the binoviewer *was* slightly dimmer than it was in a single eyepiece, but, despite this, I always thought DSOs looked better with the binoviewer than without. The effect is hard to explain, but I suspect that the fact that I was more comfortable—no squinting one of my eyes—had a lot to do with me being able to see more. Let me add that in the excellent binoviewer I tested, the combined image was not *that much* dimmer than with single eyepiece, "Cyclops style" viewing, as the binoviewing fans say. I had, for example, the best view of M37, the marvelous open cluster in Auriga, I've had in a long time. *A binoviewer helps in the city.*

Another interesting and pleasant effect of using a binoviewer is a pseudo three-dimensional (3 D) effect. Given the distance to celestial objects and the very small separation between your eyes, there is no way that you can see a DSO in 3-D as you would earthly objects. But your brain refuses to believe that. You are using two eyes, so you *must* be seeing depth. This insistence by my brain that I must be seeing in three dimensions was particularly striking with M42, the Orion Nebula, with the Denkmeier. The nebula seemed to be in the mid-ground, while some stars, the brighter ones, appeared in the foreground. Dimmer sparklers formed the "background." Completely false, of course, but a beautiful and moving effect nevertheless.

Binoviewers are *not* for everyone, however. While they offer improved images in the city and in the country—I'm convinced of that—you have to be willing to pay a premium price for this admittedly relatively small improvement. Good quality binoviewers are priced at around 600–1,000 US$ and are sold for comparable or higher prices in the UK and Europe. Believe me, you will want the highest quality binoviewer you can get. They must be precisely aligned and mechanically stable. The slightest miscollimation of a binoviewer's prisms will result in headaches and eyestrain at best and two separate images that cannot be merged into one at worst.

Your spending won't be complete when you've purchased the binoviewer, either. You'll need eyepieces for the thing. That means you will have to buy identical eyepieces for every focal length you want to use. Often you can't just buy duplicates of those you already own, either. As time passes, manufacturers change the sources and specifications of their oculars. For use in a binoviewer, the two eyepieces must be as identical as possible. If they are not, you may very well run into problems with merging images. Even if you're content to stick with inexpensive Plossls, at least for a while, setting up a binoviewer with a minimum complement of maybe three sets of eyepieces means spending real money.

The binoviewer will need a home, too, a secure home if you use it in a refractor or SCT. By "home" I mean a star diagonal. A binoviewer is a heavy piece of equipment and using one with a standard 1.25-inch diagonal may be a recipe for disaster. I once saw an expensive binoviewer drop to the ground when the barrel of the 1.25-inch star diagonal it was riding in rotated and *unscrewed* under the weight of the viewer, and the setscrew holding the binoviewer in place let go. Plan on purchasing a heavy duty 2-inch star diagonal to accommodate a binoviewer. Most binoviewers have 1.25-inch barrels, and only accommodate 1.25-inch eyepieces, but 2-inch format star diagonals are heavier duty, usually, than 1.25-inch models, and are better suited to holding a heavy accessory like a binoviewer. If possible, purchase a 2-inch diagonal that uses a compression ring to hold eyepieces in place. A compression ring not only prevents marring of the binoviewer barrel, which can be caused by tightening setscrews tightly enough to hold the bino in place, it holds this heavy accessory in place more securely.

You may even have to modify your telescope before you can use a binoviewer. Most models require a considerable amount of extra inward focus due to their system of light-folding prisms and the resulting longer light path before the image reaches the eyepieces. A moving mirror focusing scope, a Schmidt Cassegrain or Maksutov Cassegrain, won't have a problem. Some refractors and almost all Newtonians, however, will. At a minimum, you may have to insert a "relay lens" supplied by the binoviewer manufacturer between the viewer and the scope. These lenses will allow the binoviewer to reach focus, but will usually act as barlow lenses, meaning wide fields are impossible to achieve. For some scopes, even a relay lens may not work. You may have to move your Newtonian's primary mirror or shorten your refractor's tube.

Sounds like a lot of expense and trouble just to be able to use two eyes, but, in my judgment, you may still want to consider a binoviewer. I was skeptical about these devices, but after using one for a few months, I became a convert. They really do improve the observing experience in the city. The only case where I was disappointed was in objects that were at the very edge of detection in a single eyepiece. These were sometimes invisible in the binoviewer, but were no great loss. In addition, if you like to look at the Moon and planets, binoviewers can provide tremendous views, putting single eyepiece set-ups to shame.

Finding Objects in the City

All the fancy telescopes, eyepieces, and binoviewers in the world aren't worth anything if you can't find objects to view. Locating objects, dim DSOs not visible in a finderscope, is a frustrating experience for beginners under the best conditions, and in the city even experienced observers may have a hard time. Yes, those galaxies, nebulae, and star clusters *are* there, many of them, anyway, despite your sodium-pink skies, but they can be difficult to land on due to a lack of visible stars to guide you to them. But I guarantee you can find any object visible in your skies and your aperture with relative ease by using one of the following methods or a combination of them.

Brighter Objects and Richer Star Fields: Star Hopping

Star hopping is the normal method amateur astronomers use to find objects if they don't have or want to use setting circles or go-to telescopes. Star hopping is actually something of a misnomer in my opinion. What you are actually doing is making *patterns* in the stars. Say, for example, I want to find the wonderful little star cluster NGC 457. This group, known as the "ET" or "Owl" cluster due to its stick figure shape, is a good object to hop to in the city since it's located among the bright and prominent stars of Cassiopeia. How do you capture ET? Take a look at Figure 4.1.

You'll see that the cluster forms a triangle with two stars in Cassiopeia's "W," Delta and Gamma. I've indicated this with the dashed lines I've superimposed on the chart. Move your scope so that it's approximately on the place in the sky indicated and, with a little hunting around you'll have ET in your field. Do the same sort of thing with other objects. NGC 129 is about halfway along a line drawn between Gamma and Beta. NGC 663 is at the apex of the triangle formed with Epsilon and Delta. You can do this all over the sky, forming triangles, lines, and other patterns to lead you to

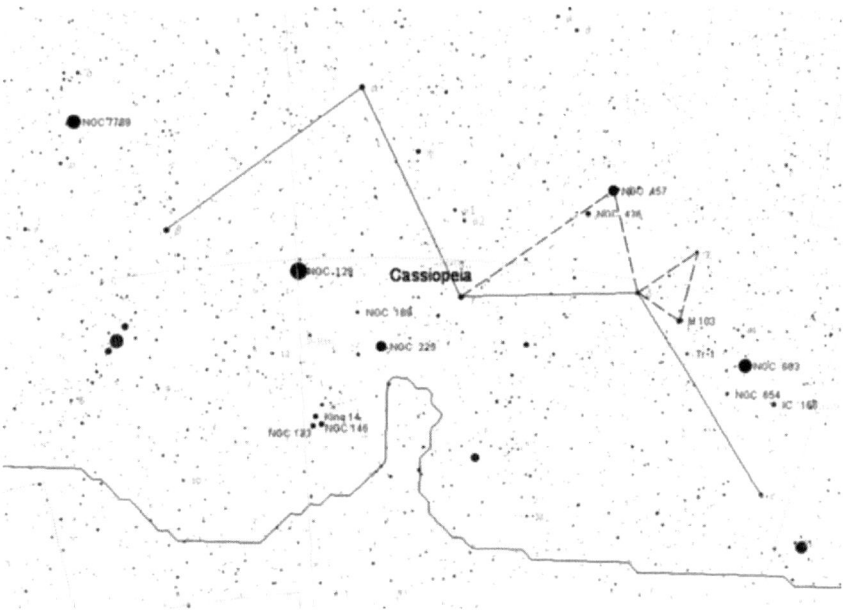

Figure 4.1. Star hopping in Cassiopeia.

objects. This works well, however, *only* if you've got enough bright stars to easily form patterns and if the object you're hunting is bright enough to show up easily with a little scanning around.

Dimmer Objects: Eyepiece Hopping

What if there aren't any bright stars in the neighborhood? Or what if the object is dim enough that you need to be more precise when looking for it—if it'll be easy to pass over while quickly slewing around? In these cases, we'll use Method 2, Eyepiece Hopping. A good area for this technique, as was mentioned in the preceding chapter, is in Virgo. Again, there are dozens of galaxies visible here from even very poor skies in medium-aperture telescopes. But if your conditions are as bad as mine, there'll be few stars visible, even in a large finder, far too few to allow you to hunt down the dim DSOs lurking there.

To begin eyepiece hopping to a target, you will have to locate a starting place by other means. This may be by star hopping—usually you can find a few prominent stars somewhere close to the area of interest to use as a jumping-off point—or setting circles. Figure 4.2 is a medium wide-angle chart of Virgo, with our starting point indicated by the eyepiece field circle labeled "1." One of the few striking asterisms visible in this area of my sky is the little Y-or arrow shaped figure of magnitude 4–6 stars near the field circle 1 is centered on. This star pattern, which includes 5th

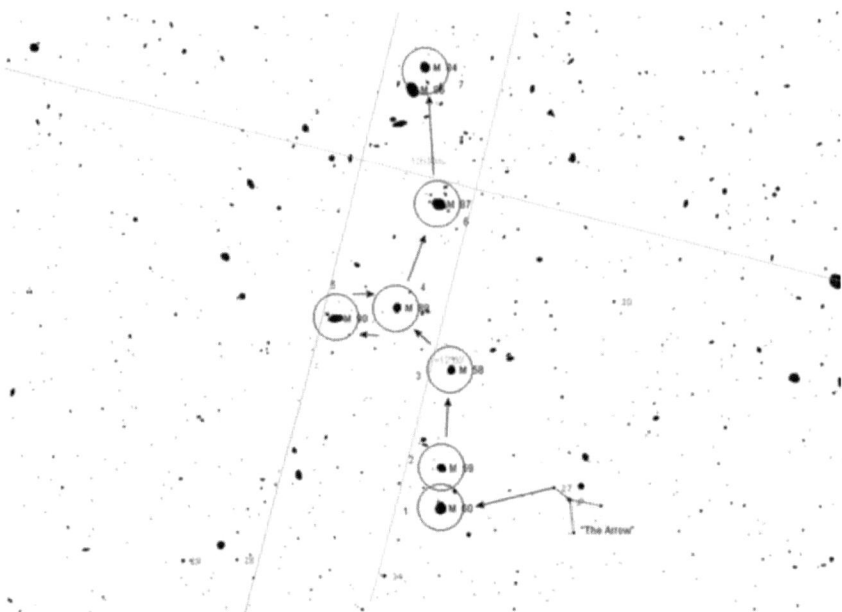

Figure 4.2. An eyepiece hop through Virgo.

magnitude Rho virginis, is easily identifiable in a 50-mm finder, and is a quick hop from bright Epsilon Virginis. Once you've got the "Y" in your eyepiece, your work is really over. If you're careful in moving your telescope, you can enjoy galaxies all night long without using your finder again.

How? Refer to Figure 4.2 again. Once you're centered on the "Y," insert an eyepiece that provides good contrast while yielding as wide a field as possible. Use your judgment and your familiarity with your own skies to decide what to use. What worked well for me was a 22-mm Panoptic in my 11-inch $f/10$ SCT. In my 12.5-inch $f/4.8$ Newtonian, a 12-mm Nagler wide-field eyepiece worked splendidly. When you're ready, while looking through the eyepiece at the stars of our little Y, move Northwest to land on the field labeled "1." How much you move depends on your scope and your eyepiece. My SCT setup meant that I had to move about 1.5 fields to the Northwest to achieve the 45′ of distance I needed to travel with my 0.5° true field eyepiece, a short enough hop to make it easy to be precise. Once you're there, look carefully. The Messier galaxy M60 will be glowing in the center of the eyepiece. What if you moved a little more to the North than you should have? That will land you on a different galaxy, M59 instead, but your mistake should be obvious when you compare the field stars to those visible in your scope.

Naturally, you'll have to spend some time planning your routes, designing them for your scopes and eyepieces, but this method is very worthwhile, and will allow you to see more in star-barren areas of the sky like Virgo than you ever thought possible (I've also used eyepiece hopping to very good effect in and around the bowl of the

Dipper/Plough). You can design eyepiece hopping charts using a detailed star atlas like *Uranometria 2000* and a pencil and compass, but I find that a computer program makes the process incredibly easier. The chart in Figure 4.2, like all the charts in this book, is based on originals done with Patrick Chevalley's free program, *Cartes du Ciel.* *Cartes* allows you to plot eyepiece field circles exactly of the correct size and orientation for a given scope and eyepiece combination, meaning that it will be clear how far you need to move to get to the next object's field. One word of advice: make your charts in a larger scale than I have in Figure 4.2, maybe placing only two or three eyepiece fields per page in order to make it easier to verify patterns of field stars to make sure you're in the right place.

The Setting Circle Alternative

When I was a young man with a new 3-inch telescope on an alt-azimuth mounting, I dreamed of getting my hands on an equatorial mount equipped with *setting circles.* Analog setting circles (Plate 19) are graduated dials, one for each axis of the telescope, with the declination circle being marked in degrees and the right ascension (RA) circle labeled with hours and minutes. The idea is simple; you move the telescope until pointers on the setting circles indicate the proper values of RA and declination for the object you're seeking (obtained from a book or chart). When the right values are dialed-in, the object of your desire should be in the telescope's field. Wow! I dreamed of cruising through the Messier list and beyond with ease, *dialing my way* across the heavens to every imaginable DSO, viewing galaxy after galaxy effortlessly. No more finders, no more star charts.

When I finally obtained a decent German equatorial with a pair of setting circles, the truth turned out to be less glamorous than the dream. For one thing, a couple of years with an alt-azimuth mounted scope had made me a proficient star hopper at age 12. I found it easy enough to find almost any object I wanted with the aid my star atlas and finder, so the effort required to learn the use of setting circles didn't seem as worthwhile. When I finally got around to figuring out how to use them, they lost even more of their appeal. Try as I might, I just couldn't locate objects with my circles. I was lucky to even get a target in the field of my finder. I forgot about setting circles and didn't use them again until I began teaching astronomy at a university and was tasked with showing students the ins and outs of analog setting circles.

Surprisingly, despite my less than happy memories of setting circles, I found they can work very well if some conditions are met. The first is that the circles must be large enough. Large diameter setting circles mean greater accuracy. On most small imported German equatorial mountings (GEMs), they are far too small to be accurate enough to put an object in the eyepiece of the main scope. For accuracy, you need something about the size of the RA circle on modern fork-mounted SCTs (8 inches in diameter). Unfortunately for SCT users, the declination circles on these scopes approach too small at about 4 inches, but even that is much better than what you'll see on most imported mountings.

Accurate setting circle *calibration* is another requirement. Declination circles should be properly set up at the factory, but they do occasionally require readjustment. When a fork-mounted scope's tube is parallel to the fork arms (determined by the use of a bubble level) or a GEM's tube is precisely parallel to the RA shaft, the declination circle must read *exactly* 90°. The mount's RA circle must be calibrated every single night, as it will lose its alignment as soon as your clock drive is turned off—assuming it is driven by the scope motor, an important plus for any scope. A "driven" RA circle maintains the same reading as the scope moves to track the sky. An undriven circle slowly loses accuracy, becoming a little more "off" as the scope tracks. If the RA circle is not driven, you'll have to continually reset it all night long. To calibrate the RA circle, driven or not, place a star of known right ascension in the field of a high-power eyepiece and adjust the circle till it reads the RA value of the star.

The final, and possibly most critical, requirement for satisfactory setting circle performance is accurate polar alignment. In order to locate objects, especially those far from the star you use to align the RA circle, your scope's RA axis must be pointed very close to the true celestial pole. The better your polar alignment, the better your results will be. Of course, you reach a point of diminishing returns, and I don't feel it's necessary to do a photographic-precision drift alignment to use setting circles. On the other hand, just pointing the scope's RA axis at Polaris will rarely yield satisfying results. A good compromise is a polar alignment scope. Either a bore-scope built into a GEM's right ascension axis, or, for a fork-mount scope, a standard finder with a special polar alignment reticle. These devices will yield an alignment "good enough" for setting circle usage without requiring so much time for an alignment that you'll be reluctant to do it.

By scrupulously observing all the above requirements, I've had very good success with my SCTs' circles. I don't always get an object in the field of even a low-power/wide-field eyepiece, but the quarry I'm hunting is usually within half a degree of where I land, and all I have to do is to carefully and slowly slew around a little bit to move it into my field.

What if you're the owner of a mount with too small circles? Or your skies are so bad that it would be a big help to get dim objects in the field of a low-power eyepiece every time? If an object is on the very limit of visibility in your skies, hunting around blindly over comparatively large areas may not always yield results. If you *know* the object is in the field of your low-power eyepiece, you can up the power to increase your contrast, and usually capture it with just a little hunting around. To get this increased accuracy, though, it's necessary to forget analog circles and go digital.

Digital Setting Circles

When digital setting circles (DSCs) first became popular with amateurs in the 1980s, that's all they were, "digital setting circles." They merely replaced your analog-readout circles with a digital display. Your scope had to be precisely polar aligned to make use of them and, while accuracy may have been slightly better than with analog circles due to their finer readouts—they solved the "too small circle" problem—accuracy was generally similar to what a careful user of quality analog circles could achieve. Then the computer revolution hit amateur astronomy full-force.

By the early 1990s, DSCs had become real computers and quickly made analog circles obsolete for observers seeking high accuracy. The most important advance was that the new DSCs made precise polar alignment less important. Many DSC computers, then and now, work better with an accurately polar-aligned telescope, but can deliver very good accuracy with nonpolar aligned instruments, including Dobsonians. About this same time, DSCs also began to offer large internal databases of objects. You no longer had to look up coordinates and match them to the DSC display. Want to go to M13? The readout of the newer circles indicates which way to move the telescope and tells you when to stop.

An example of a modern digital-setting circle computer is shown in Plate 20. This is the Argo Navis DSC system, which offers many of the same features as the most advanced go-to telescopes. A unit like the Argo Navis can actually make a "full go-to" instrument unnecessary for many observers.

What can DSCs do for you? With an equatorial telescope, they can deliver consistent 0.5° accuracy across the sky, putting objects in the field of a low-power eyepiece every time. They can do just about as well with a Dobsonian, assuming that its rocker box is solidly built and square—"orthogonal." In order to achieve this accuracy, a few conditions do have to be met. The encoders, the little optical sensor units that attach to each axis of the mount and tell the computer where it's going and how fast, must be carefully installed. If the shafts of these units are not firmly connected to the scope mount (by means of gears or belts, usually) the encoders will slip and the computer will be misinformed about its position. Care must be taken with the two alignment stars used to set the DSCs up at the beginning of a session, too. Follow instructions given by the DSC maker about choosing stars (usually you'll want stars about 90° apart in the sky, no more, no less), and use a high-power eyepiece when performing the alignment.

Are DSCs a good buy for the urban observer? Maybe. If you've already got a telescope that's well suited for city lights observing, the answer is a definite "yes." Though top of the line DSCs are relatively expensive (around 500 dollars or pounds), they can go a long, long way to making your viewing more productive and more fun. On the other hand, if you don't yet own an instrument suitable for city use, you might want to consider a go-to telescope instead.

Go-To for the Urban Astronomer

Go-to telescopes, computerized scopes that point themselves to desired objects at the press of a few buttons, have been sold to amateurs for almost 20 years now. Celestron in the U.S. offered the first mass-produced go-to SCTs, the Compustars, in the late 1980s, and its competitor, Meade, brought-out the first "popularly priced" computerized SCTs, the LX200s, in 1992. Go-to telescopes have come a long way in 20 years. The initial scopes were not really more accurate than DSCs, had small object libraries, and were offered only in the SCT optical design. Now, go-to scopes, even those with rather large apertures, cost far less in real currency than the early models did. They are also equipped with libraries of DSOs ranging up to 100,000 objects (though you're unlikely to see many of these with an 8- or 10-inch telescope). Go-to has also moved far beyond SCTs. The go-to Schmidt is still probably the most common computerized scope, but it is not at all uncommon to see an APO refractor or Newtonian pointing itself at DSOs anymore.

Should you buy a go-to telescope for city observing? In practical terms, in terms of reliability and utility, the answer is YES. Today's go-to instruments are considerably more reliable than the old Compustars or the early LX200s. Most importantly, they are a boon for observers working under light-polluted skies. If you can see two alignment stars, you can get any object you choose in the field of your scope. Being able to see the DSO you are after is not always a given, but it will be there. Current go-to telescopes are usually considerably more accurate than DSCs, with pointing errors often being measured in just a few arc *minutes*. My current personal scope, a Celestron Nexstar 11 GPS, will, for example, consistently put anything I ask for in the field of a TeleVue Nagler 12-mm at 220×, which pulls many, many objects out of the bright sky. I think go-to is the greatest thing to ever hit urban astronomy.

Not everybody thinks go-to is such a good thing, however. If you read the Internet astronomy message boards or hvae listened in on a heated conversation or two at your local astronomy club, you're aware that some observers are opposed to go-to on *philosophical* grounds. They claim that by automating the finding process amateur astronomers, particularly new amateur astronomers, miss out on the fun that comes with learning the constellations and star hopping. I wish some of these die-hard traditionalists would pay a visit to my backyard on a spring night and have some "fun" star hopping through Virgo. I'm willing to bet they'd lose heart in that bright desert soon enough and make friends with my Nexstar 11.

I do think amateurs should learn the constellations as a matter of personal pride, but telling an urban observer to star hop doesn't have the desired result of making him or her a better observer. It often just means she'll drop out of the hobby due to the difficulty inherent in star hopping in city skies and the boredom of not seeing much. *If you observe in the city, get a go-to telescope if at all possible.*

A question I'm often asked is, "Can I retrofit my manual telescope for go-to?" The answer is "maybe." If you have a fork-mounted telescope—SCT, MCT, or other—the answer is no. Installing go-to on a fork would require major and extensive modifications. Talented individuals have done custom go-to mods for their fork-mounted scopes, but this task is far beyond most of us. If you have a German mount, though, the answer is a qualified "yes." Various manufacturers offer go-to drives for a variety of GEMs, and especially the Japanese Vixen Great and Super Polaris and their "clones," the Chinese EQ4s and CG5s. The Losmandy G11 and GM8 GEMs can also be given go-to capabilities.

Most of these add-on go-to units, which are available from a variety of manufacturers including Vixen, work very well. Go-to German mounts also provide an upgrade path for fork-mounted SCT owners. You can't equip the fork for go-to, but if you have a good optical tube, you can remove it from the fork, put it on a go-to GEM and keep on truckin'!

The scope is ready outside, you've gathered your accessories, and are prepared for an evening of city observing. What do you observe? Which objects look best in the city? Which are impossible? How do you tell the difference? That's the subject of the following Chapter 5.

CHAPTER FIVE

Urban Observing Programs

On any evening under the stars, a few wonderful deep sky objects (DSOs) draw me to them like a moth to flame. These are the beautiful and bright deep sky showpieces like M13, the Hercules Cluster, and M42, the Great Nebula in Orion. M13, M42, and the other bright and outstanding Messiers deserve repeated observation, certainly, but eventually you'll have to move beyond the "best of the best" or risk boredom and a loss of interest in observing. Also, just observing the bright easy ones doesn't help hone your observing skills, one of the reasons we gave in Chapter 1 for observing from the city.

I find that having an observing project, a specific program of challenging and new objects, helps me move beyond the best Messiers, see more, and keeps me interested in observing every clear night. If I have a project I'm working on, I'm more likely to get outside with the telescope than I would if I were just observing "randomly." Having a pre-prepared list of objects to view also means I don't waste time trying to figure out what to look at next—if I don't have a list, I tend to do a few of the brightest objects available on any given evening and call it a night.

What kind of observing project? When it comes to the deep sky, in the city as well as in the country, the place to start *is* the Messier list. Most of these well-loved objects are easy pickings for careful astronomers in light-polluted areas, and they give you a good idea of what you can expect from the deep sky from your location before you move on to more difficult challenges. For that reason, start with the Messiers even if you've viewed the entire list of 110 objects before from better observing sites. Don't just stay with the M13s and M42s, though. The Messier contains some frustratingly hard objects that you should make it a matter of personal pride to view from the city.

The Messier: Tough and Easy

Sure, M13, M5, M42, M45, and quite a few others are genuinely bright and easy for small urban scopes, but there are some genuine toughies, too. The hard ones fall into two categories: bright and dim face-on spiral galaxies and challengingly dim nebulae of all kinds. In the former category are those legendary island universes, M101, M74, M33, and similar galaxies. The latter group includes M97 (the Owl Nebula), M76 (The Little Dumbbell Nebula), M1 (the Crab Nebula), and a few others. All of these objects can be very difficult to see from an urban site on most nights, but don't despair if you've just hunted fruitlessly for M74 for the 10th time. It *can* be done. You just need preparation and knowledge on your side.

Preparing to Observe the Messier

Don't just haul the scope out into the yard, glance at a copy of the Messier list and start hunting. That is a recipe for frustration. Oh, you'll have some successes, but you'll have plenty of failures, too. To minimize these failures you'll need to ready yourself by knowing the "when" and "what" of urban observing. "When" as in "when do I look for a particular object?" and "what" as in "what will it look like when I find it?"

When to Look for Deep Sky Objects

You probably already know you should wait until an object is at least 30° above the horizon before trying to find or observe it. This is true in both country and city, since, even in the country, the thick air and dust near the horizon can really dim-down a galaxy, nebula, or star cluster, but this is an especially important consideration in the city, as light pollution will be at its worst closest to the horizon. In fact, you may find that you have to wait until the area of interest is considerably higher than 30° before it clears the heaviest sky glow, depending on your site. You may even have to hold off until the object is in a different part of the sky altogether. My home skies, for example, are brightest to the east, with the west being noticeably darker. It helps, then, if I wait for an object to move well away from the eastern sky before looking for it. There is always a best time to view any DSO regardless of the particular characteristics of your site: the time of its *culmination*.

Culmination: As Good as It Gets

To understand "culmination," you have to understand the term "Local Meridian." The Local Meridian or "Meridian" is an imaginary line that runs from the North Celestial Pole, though the Zenith (the point directly over your head), through the South Celestial Pole, through the Nadir (the point directly beneath your feet), and back to the North Pole. The Local Meridian never moves. As the night wears on, stars and DSOs (and the Moon and planets) rise in the east, move across the sky, cross the Local Meridian, and set in the west. By definition, then, when a star, a DSO, or anything else is on the Local Meridian, it is as high in the sky as it will ever get. Depending on an object's declination north or south, that may not be very high. When an object with a declination that places it far north or south of the Celestial Equator is on the Local Meridian, it is still not very high above the northern or southern horizon, but the moment when it is on the Meridian is *still* as good as it gets for that object elevation-wise. The moment when an object hits the Meridian is "*culmination.*"

It is important for urban astronomers to know when this will happen, since the time when a DSO culminates is the time when it's farthest away from the dust, thick air, and sky glow near the horizon. Note that objects close to the celestial pole can culminate below the pole as well as above it. We're interested in the time when they culminate above the pole so they are away from the horizon as much as possible.

Planispheres, Computers, and Local Sidereal Time

How can you tell when an object of interest, say, galaxy M101, will culminate so you'll have a prayer of seeing it? An uncomplicated way is to use a Planisphere, one of those little paper or plastic "star wheels" that can be rotated to show how the stars will look at a given time and date. Rotate the wheel until M101 is on the meridian. Most planispheres will not have a marked line for the local meridian, since it has to remain still while the sky rotates beneath it, so imagine a line extending from the rivet that holds the sky-disk in place on your planisphere and running across the sky to the point marked "12 Noon" at the bottom (Plate 21). You can then read the planisphere's date/time scale to determine when M101 will be the highest in the sky. Naturally, if you've got a computer running a planetarium program, it's easy to change the sky until M101 is on the meridian and then read the time for your location.

Local Sidereal Time

Another way of determining when a DSO will be on the meridian is by using Local Sidereal Time (LST). Local Sidereal Time is easy to understand. Every object in the sky, as you're probably aware, has a Right Ascension (RA) value given in hours, minutes, and seconds. Local Sidereal Time is the value of the line of RA currently culminating on the meridian. M101's RA, which you can look up in a book or on a chart, is 14 h 03 min. When M101 is on the meridian, it's 14:03 LST.

With a special clock that reads sidereal time and a star chart at hand, we can both determine what's on the meridian now and what will be on the meridian at a given time. For example, if it's 12:03 now (in LST), it's easy to see that M101 will culminate in approximately 2 h. But how do you determine what the LST is? Formerly, you had to have a special clock, one whose day is 23 h, 56 min, and 4 sec in length rather than the 24 h of a Solar day (due to the Earth's movement in its orbit, the Solar Day is longer than the Sidereal Day). With the coming of the computer revolution to astronomy, anyone can now have a sidereal clock for little or no cost. Sidereal clock programs are available for both PCs and pocket computers like the Palm and the Pocket PC. Most planetarium programs will also provide a read-out of LST.

What Will the DSO Look Like? Surface Brightness

Finally, a good night in the city! Joe and Jane Newamateur are not about to waste it, either. They've started on the Messier list, had some initial successes, and intend to press on on this crisp and relatively dry spring evening. What to hunt for? Obviously, galaxies. If you're looking to the east, there's not a whole lot else to attract your attention during the spring of the year other than the rather lackluster globular M53. *Which* galaxies? Jane feels she's prepared well for this session. She has a notebook full of lists of objects printed with her deep-sky planning software, and has sorted the entries according to type and constellation. Looking thorough Ursa Major's entries, she comes to M101. "Here's a good one, Joe. M101. It's pretty bright, magnitude 7.9."

Joe and Jane start out bravely. At magnitude 7.9, this lovely face-on Sc spiral ought to be easy pickings for their 8-inch Dobsonian, but after half an hour of fruitless searching, they give up. This "bright" galaxy is nowhere to be seen despite the fact that they've star hopped carefully. "Oh, well. How about M97, the Owl Nebula?" It has an interesting name and looks fascinating in the pictures they've seen in deep sky guides. Its stated visual magnitude is a little dimmer than that of M101, but it should be easily visible in their dob, they think. Half an hour looking for M97 yields nothing. Joe and Jane move the scope back inside, dejected about their puzzling failures with these supposedly bright objects.

If a slightly more experienced urban astronomer had been on the scene, she could no doubt have told Joe and Jane that M101 and M97 are hardly easy from the city. Or even from the country at times. They are both, and especially M101, a face-on Sc galaxy, very subdued. But why? The catalogs clearly state that they are 7.9 and 9.9. The reason is that these stated magnitudes are the total *integrated* brightnesses of these DSOs. 7.9 *is* the magnitude of M101, but it assumes that this big 22 arc-minute-across spiral has been *compacted* to the size of a star. Find a magnitude 8 star and defocus your telescope until it fills a field of view 22 min across, and you'll soon get the idea. As the star expands, it becomes dreadfully dim. The key to the true brightness of a DSO is its *size*. The bigger a DSO is, the brighter it has to be to show up well.

How do you find the *true* brightness of a DSO? Many computer programs will give this true or "surface" brightness for DSOs (often listed as magnitude per square arc second or minute) as well as the integrated magnitude, how bright the object would be if squished down to a star-like point, which is usually labeled as the "V" ("visual") magnitude. With a little experience, you'll soon be able to get a fairly accurate idea of how really bright an object will be just by glancing at its size and

V magnitude. For example, it's easy enough to see that M33, with a size of 73 arc minutes by 45 arc minutes, is going to be very dim, even at its bright V magnitude of 5.7.

How can you figure true magnitudes, the true *surface brightness* (SBr) of DSOs yourself? A simple formula can be stated as follows:

$$SBr = m + 5 \times \log(d) - 5 \times \log(70)$$

where

$$SBr = \text{surface brightness;}$$
$$m = \text{integrated (visual or V) magnitude;}$$
$$d = \text{size of object in arc seconds.}$$

Using this formula, we find that the supposedly "bright" M101 has a "true" magnitude of 14.2, which makes it *quite* a challenge in the city. At V magnitude 9.9 and a size of 3.4 arc minutes across, M97 turns out to be at 13.63. Both of these objects are quite doable from the city with the right techniques and equipment despite these off-putting figures, but they are nowhere *near* as easy as their simple visual integrated magnitudes would suggest. The above formula, since it allows you to enter only one dimension of the object, works best for round DSOs, and will not give an accurate figure for a long, thin galaxy, for example. In my experience, though, the big round ones are the hard ones, in the Messier list. Among the Messiers, it doesn't get more difficult than face-on spiral galaxies and big planetary nebulae, both more-or-less round-shaped objects.

Aside from surface brightness issues, the size of an object can affect your ability to see it in another way. A planetary nebula can be relatively easy, even if it's large like the Owl, if it can be *well framed* in a medium-power eyepiece. If you can increase the power enough to darken the sky background, but still leave enough background *around* the object to provide a visible contrast difference. Consider M33. At 73 arc minutes across its major axis, it will totally fill the field of a medium-power ocular. All you'll have in the field will be galaxy; there will be no sky background to *compare* it to, and it will be *very* hard to see. Naturally, you can reduce power until the galaxy is better framed, but the Catch 22 in the city is that by doing so you brighten up the sky background and there *still* isn't enough contrast to pop the object out. Bottom line? Bigger equals harder.

Smaller isn't always easy, but it sure helps. Small planetary nebulae are a natural for the urban observer. With these objects, surface brightness is on your side. At approximately 1′ (arc minute) across, the Ring Nebula, M57, for example, really does look almost as bright as its published integrated magnitude of 8.8. In a 4-inch scope it can be easily seen as a delightful little smoke ring. There are countless other planetaries scattered across the Milky Way that are in the 1′ or smaller size range. With these small nebulae, you can push even a smaller scope farther than you might think, going to magnitude 10 or 11 planetaries with a 4-inch scope. At a certain point, however, as planetary nebulae get smaller, things get hard again.

The exact *small* size where they get tough will depend on the observer, telescope and magnification, but at less than 5″ (arc seconds) in diameter, planetary nebulae become difficult to distinguish from stars. You can increase the magnification in hopes of making these little things look non-stellar, but on nights of poor seeing this may increase the apparent sizes of field stars too. One way of dealing with small planetaries is to look for a blue–green color. Planetary nebulae often emit light of this color because they are radiating at the wavelength of OIII due to doubly ionized oxygen in their makeup. Unfortunately, not all planetaries look blue–green, no matter how large your scope is. Quite a few just look gray. And the dimmer they are, the harder it is to detect color.

One clever method of picking out these objects is to "blink" them with an OIII filter. Theoretically, if you compare the view with an OIII filter to that without, the nebula should be more visible with the filter in place. If you switch the filter in and out rapidly by passing it in front of your eyepiece's eye lens, it should appear to *blink* in the field. Naturally, the planetary must respond to an OIII filter well— something not all of them do—for this trick to work. Both methods of locating tiny planetaries can be supplemented by a comparison of what you see in the field to a detailed chart made with a computer program. A combination of these methods will almost always unmask even the smallest and most stellar-appearing planetary nebulae.

Urban Observing Projects Beyond the Messier

Yes, Charles Messier's famous list is the obvious place to start if you're looking for an observing project. But what happens when you complete it? What if you're struggling with the last few difficult objects of the Messier list, are waiting for an especially good night to track them down, and want another project to work on while hoping for a superior evening? Certainly you can take on the objects in this book, but you'll eventually exhaust them too. What next? A good place to look for ideas is the Astronomical League's website (see Appendix 1 for the URL). The Astronomical League is the U.S. umbrella organization for both individual amateurs and astronomy clubs. One of its most popular functions is running *observing clubs*. Complete a list of objects and you'll be awarded a nice certificate. The AL has a variety of these "clubs" ranging in difficulty from the Messier to the somewhat forbidding Herschel 400 II (a list that contains some truly difficult objects). The Herschel II might be a bit much if you're in the worst of the light pollution, but one new AL club is a natural for you, the Urban Club.

To qualify for membership in the Urban Club, an amateur astronomer must view at least 100 DSOs from a list found on the League's web site. You must observe these objects from a suitably light-polluted area (this is the only time I recall seeing light-pollution listed as a *requirement* for deep sky observing), which is defined as any area where the Milky Way is not visible. To qualify for the club award certificate on completion of the list, you must be a member of the Astronomical League, but anyone, anywhere can enjoy working through the League's well-thought-out Urban List.

There's also a second Urban List you can pursue, one composed of double and variable stars. Many observers say that they prefer "real" DSOs to stars, but doubles, in particular, can be incredibly beautiful and fun to observe.

Your Own Projects

There are numerous other lists available online and in the astronomy magazines and books for you to try as your "next" project. Unfortunately, though, most of these are not aimed at city observers. When you run out of ready-made observing programs for the city, what do you do? You make your own. The actual construction of an observing list is made easy with programs like *Skytools 2*. Or you can just use a pencil and a piece of paper. Whichever method you choose for putting together observing lists (I prefer computer programs since you can easily sort objects as desired), you'll need a source of data on DSOs to help you decide what to put in your list and what should be left out. *Skytools, Deepsky,* and other astronomy software will give you some information about the DSOs in their databases, but this may not always be in-depth enough to allow you to decide whether that 11th magnitude galaxy should go on your current city list or not. Most amateurs turn to one of two sources of printed information on the deep sky for help in choosing objects, *Burnham's Celestial Handbook* or *The Night Sky Observer's Guide.*

Burnhnam's

Burnham's Celestial Handbook by the late Robert Burnham Jr., first published in its present form in 1976, is a wonderful and beautiful book in every way. Composed of three fat volumes, the *Handbook* covers the entire sky from celestial pole to celestial pole. Its data is arranged by constellation, one chapter per constellation, with each chapter containing a list of interesting sights—deep sky and stellar—and detailed descriptions of the constellation's most prominent objects. The information given about the important objects is more than detailed enough to allow you to determine what the subject will be like in your urban environment.

Unfortunately, once it was published, Burnham's was never updated in any meaningful way. This mostly affects the "theoretical" information contained in the book's introductory section is dated—obviously science has come a long way in the last 30 years in understanding how stars, galaxies, and the universe work. This outmoded data won't cause a problem for observers. What *may* trouble the active astronomer is the coordinates in Burnham's. The right ascension and declination values are all given for Epoch 1950.0. Due to precession, the wobble of the Earth's axis, the R.A. and declination coordinates of DSOs have changed significantly from their 1950-calculated positions. Again, this shouldn't really be much of a problem for the practical astronomer. It's easy enough to find an object on a chart and determine its epoch 2000 or later coordinates if necessary. There are even small freeware computer programs available that will "precess" 1950 coordinates to 2000—or whichever epoch you desire.

What keeps me coming back to Burnham's? Not just the hard data, though there's enough of that in the book to last most astronomers—city or country—for a lifetime. It's the beauty of Burnham's prose. His thoughtful, aesthetic take on the heavens is something every amateur astronomer should experience. For example, he doesn't just state the bare facts on how globular cluster M22 looks—big, bright, and resolved in small telescopes. He goes on to say how it reminds him of a passage in J.R.R. Tolkien's

Lord of the Rings. He doesn't just rattle off Sirius' vital statistics; he prints a poem about the Dog Star. No matter how the heavens precess, or how many generations of amateur astronomers come and go, *Burnham's* will always be loved.

The Night Sky Observer's Guide

Just because *Burnham's* is a *good* thing, that doesn't mean it's the *only* thing. Many amateurs will tell you the recent book, *The Night Sky Observer's Guide* by Robert Kepple and Glen Sanner, is a must-have for any DSO observer. *Sky and Telescope* magazine described this as "*Burnham's Celestial Handbook*: The Next Generation." Well, not quite. This book isn't the next Burnham's, and it doesn't try to be. *Burnham's* was a special work that's not easily duplicated. What deep sky observers wanted was not really another Burnham's, anyway, but a book on deep sky observing that contained many more objects than Burnham's with more details about their visual appearances and Epoch 2000 coordinates.

The Night Sky Observer's Guide (*NSOG*) delivers this—in spades. Like Burnham's it's arranged in constellation order. Unlike the previous deep sky bible, though, *NSOG* dispenses with Robert Burnham's philosophical and cultural references. You do get a short description of the mythological background of a chapter's constellation, but after that it's all hard object data, finder charts, and descriptions by visual observers.

Of particular value for urban observers is that the authors almost always include data on how a given object looks for a particular aperture range of scope. You'll get details, for example, on a galaxy's appearance in a 6–8-inch scope, a 10–12-inch one, and a 16-inch or larger instrument. If an object is overly difficult or impossible with a certain aperture, there will, naturally, not be an entry for that aperture. This makes choosing objects for an urban list very easy. If I'm using an 8-inch instrument in the city, for example, and I see that a galaxy was dim under the good conditions used for observations in this book, or if it's not described for an 8-inch scope at all, I usually move on to something else. In addition to its numerous photographs of DSOs, *NSOG* also features drawings for many DSOs. I like that, since I can usually get a better idea how particular target will look visually from drawings by fellow amateurs rather than long exposure photos.

Other Information Sources

There are numerous other deep sky observing books, but one standout is the series of books by the Webb Society (a membership organization for deep sky observers) in the UK. These volumes cover the full range of DSOs in 8 volumes, with each book devoted to one class of object. The books, while not copiously illustrated, are well written and an excellent reference for any deep sky observer. Unfortunately, the only volumes in the series still in print are the final three: *Anonymous Galaxies, The Southern Sky,* and *Variable Stars.* Nice books, but a little less interesting for the general DSO fan that the earlier volumes that covered meat and potatoes DSOs like galaxies and globular clusters. You may be able to find the earlier books from used sources.

Online Information

Before the Internet, back in the medieval days of the 1970s, the data available to the average deep sky observer generally came from two sources, *Burnham's* and the monthly columns in *Sky and Telescope* by that Dean of deep sky observers, Walter Scott "Scotty" Houston. We've lost Scotty now, and every DSO observer of my generation misses his steady, knowledgeable guidance. Our loss is made up for—at least in part—by the explosion of the Internet. Turn on your computer and you can browse deep sky databases for weeks. Everything from NASA's Extragalactic Database to the huge Principal Catalog of Galaxies (PGC) is available for any amateur to peruse.

For the novice, however, most of this is a bit overwhelming. Where to start? One good place is *SEDS*, the website of the organization "Students for the Exploration and Development of Space." Their web pages include easy to use illustrated databases of both the Messier and NGC. Also very helpful is *Skyhound*, owned by *Skytools* 2 developer Greg Crinklaw. *Skyhound* is packed with information about a wide range of DSOs. One of my personal WWW favorites is *Adventures in Deep Space*. This page, subtitled "Challenging Observing Projects for Amateur Astronomers," is aimed at advanced observers blessed with dark skies, but there is still a lot here to interest the urban astronomer. You'll find the web addresses for these pages in Appendix 1.

Your Own Tours

You've got the information, now what do you *do* with it? Once you have the data, the rest is easy when it comes to producing your own lengthy projects or single-evening "sky tours." What you put on your observing lists is strictly up to you, but here's what *I* do. I generally limit myself to just one or two constellations per list, constellations that will be well placed for viewing—at or near culmination and out of the worst sky glow—on a given evening. Staying within a relatively small area means you'll develop intimate knowledge of that part of the sky, always a good thing, and it will help you focus on a constellation's less prominent treasures. In order to fill an evening's observing hours, you'll have to try for the harder stuff rather than just zipping from M13 to M57.

In selecting objects, I also try to choose a good variety when it comes to types. Just looking at one thing—galaxies, for example—can become a bit tiring after a while. I try to assemble a list with some variety; one that includes at least a couple of different species of DSOs. Naturally, in some parts of the sky—Virgo, for example—there is a preponderance of one class of DSO, and it may be difficult to find candidate objects of other types. No matter which constellation I'm "working," I always try to include a few objects challenging enough that I'm not sure I'll be able to see them at all. Growing as a deep sky observer, especially in the city, means pushing yourself.

Finally, *I save the good stuff for last.* The showpieces, the M57s and the M22s, are wonderful anytime, but looking at these first, before the harder stuff and the objects I've never seen before, always seems anticlimactic. Once I've worked through the more difficult objects on my list, it's wonderful to end the evening on a high note with a blazing Messier globular cluster or nebula. Even if I've had many failures with the earlier objects on the evening's list, viewing an easy showpiece object or two as a finale means I always end my observing run on a high, satisfying note!

Recording Your Observations

Just as some dark sky observers find it humorous that I go to so much trouble planning my bright-sky observing expeditions, some also find it funny that I'd record what I see in a logbook and make drawings of many of the objects I observe. "What's to draw?" they ask. "A lot," is the answer. Once you take your first steps in city observing, you'll quickly see that many objects present a wealth of details worthy of recording either in prose or as a drawing. Even if there *is* enough detail in urban objects to justify drawing and describing, why bother? For a couple of reasons. First, because drawing or writing a verbal description of an object in a logbook makes you work to see every possible detail in an object. Habitually drawing and logging objects fine-tunes your observing skills. After faithfully keeping a log for a while, you'll find you've become a *far* better observer than you were before. Also, these logs make up an historical record of your observing life. In the future they'll be a source of pride, a series of fond memories, and a useful tool—whether viewing from the country or city, I find it invaluable to look back on past log entries for an object. Being able to recall what exactly I saw helps me see more the next time.

How do you keep a log of drawings and descriptions? It can be as simple as a spiral-bound notebook, full of sketches and impressions. If you want to be a little more formal, you can use a computer to make up an observing form. Mine is shown in Figure 5.1. In addition to space for an adequate description, I've got entries for time, date, seeing conditions, and other variables. I like to keep my drawings and logs together, so I add an eyepiece "field circle" for this purpose. I designed a separate form for each type of object, but one version could probably serve for all of the denizens of the deep sky zoo.

Actually, I've stopped using a physical logbook. I still use my form for making drawings and taking notes at the telescope, but the final logbook these entries go into is a *virtual* one. I've finally converted my handwritten log entries into electronic form with the aid of *Skytools 2*. In addition to the planning features mentioned earlier, this program provides an electronic logbook. You can even scan in your drawings and append them to log pages. I make a sketch and notes on my original form, but transfer them to the computer the next morning. If you use a laptop computer in the field, you can make notes directly into the program, though you'll still likely want to make your drawing the old fashioned way—with pen and pencil. Most amateurs take to keeping a log right away. There's no wrong way to do it, after all. It can be as simple as a few hurried notes about what a nebula looked like. Or it can be as elaborate as an "astronomical diary," including your subjective impressions about an object and details of the urban setting you're observing from in addition to bare object details.

Even people who get very creative with their logs can be reluctant to take up a sketchpad. "I can't draw," they say. Actually, just about anybody can be taught to draw

GALAXY
OBSERVATION SHEET

OBSERVING RUN: 123 / 1

CATALOG NO.: NGC 2903   CONSTELLATION: Leo

DATE: 3/11/96   TIME: 0114   SEEING(1-10): 6

TELESCOPE: 12.5" f4.8   OCULAR(S): 12 mm Nagler II

Questions to be answered in "Description."

1) Can this galaxy be seen with direct vision or is averted vision required?
2) What is the overall shape of this galaxy?
3) Is a core noticeable? Is it compact or stellar?
4) Are the edges of the outer envelope sharp or diffuse?
5) Can any detail or mottling be seen in the outer envelope?
6) Are there any other deep sky objects in the field? What are their names?

DESCRIPTION:
(CONTINUE DESCRIPTION ON REVERSE IF NEC.)

Under these skies, NGC 2903 is an ill-defined, but easily visible glow; Seems elongated. What I took to be a stellar core may be a field star.

Field Drawing & Notes

125 X

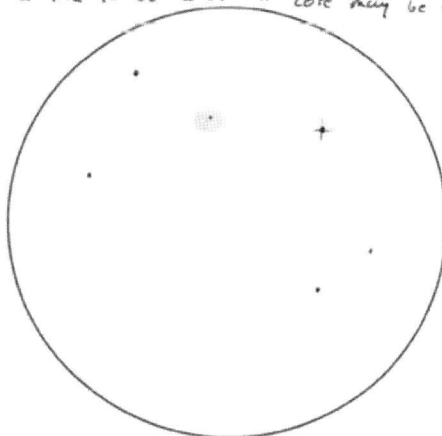

Ocular= 12 mm Nagler 2

Figure 5.1. A page from Rod Mollise's logbook.

---

anything given a little time and a good teacher. But there's very little to teach when it comes to drawing the deep sky. I can offer a few guidelines to get you started in this simplest form of astronomical "imaging," however.

The supplies you'll need are few and inexpensive and are available from any shop selling art supplies. Your basic drawing tools will be pencils. Choose an assortment ranging from hard—"2H" is good for making small stars—to some soft "nebula pencils," easily smudgeable ones like a "4B." If you want to draw on real drawing paper, choose something like a spiral-bound sketch diary. Make sure each page is large enough to accommodate notes as well as drawings. You'll also need an eraser, with an art-gum type being best for our purposes. If you're using a sketchpad or book for your work, take a compass and draw some eyepiece fields on the pages, leaving plenty of room for written notes.

But how do you *do* it? How do you draw something as confusingly complicated as the Orion Nebula? Even in the city, it displays so many details as to make it hard to know where to start. Easy. You make your drawings in *stages*. Begin with the brighter field stars. Using a medium-hard pencil or a pen (I like to use a black marker-type pen for the brightest stars) place these stars on the "eyepiece circle" on your paper. Take care to situate them accurately. Indicate brighter and dimmer stars with larger and smaller dots. Next, move on to the dimmer stars, again using the pen or a hard pencil. With the stars in place, take a good look at the nebula. Try to determine how it lies in relationship to the stars. Then, take a deep breath and start lightly shading it in with a soft pencil. Aim for accuracy, and take notes if you need to: "nebula fades into blackness here." Don't try to be overly artistic, just get the basic shapes and tones down. Use a variety of magnifications: low power to get a good overall impression of the object, and higher powers when you're hunting details.

When you look at your drawing in the morning, you'll probably be disappointed. Trying to juggle sketchpad, pencil, and red flashlight while peering through an eyepiece, tracking the object if you've got a dob or other unmotorized telescope, and trying to keep the city's ambient light out of your eyes does not lend itself to beautiful artwork. If you've worked carefully, however, you should have enough written and drawn "notes" to allow you to create an attractive finished drawing.

Start a second drawing, which can be either on the same page or a new page of your log, trying to place stars and nebulae in exactly the same places as you did in the original. By the light of day you can make the drawing look much better by doing simple things like smudging the shaded-in nebulosity with a finger. Once you're done, take one last look at your original drawing and notes, and search your memory for last night's impressions, adding any details you think will improve the drawing. Resist the temptation to put in things you've only seen in photos, though—you want this to be an accurate record of what *you saw with your own eyes*. When you're done, you'll have a wonderful record of your view. An example of my own style of drawing is shown in Figure 5.2. "Artistic"? Hardly, but it records what I saw on a given night with a particular telescope.

Accuracy is important, but there can be times when you have to *indicate* impressions rather than exact details. For example, it's easy enough to draw-in every star in even rich open clusters, but trying to record every single star you see in a globular cluster like M13 can be frustrating unless you're using a very small scope. Some observers do try to accurately draw every star they see in M13 and other bright globs, but it's generally good enough just to put in enough to indicate broad impressions. While we're striving for accuracy, this is still an *art*. You shouldn't try to imitate a camera; instead convey the proper *impression* of what M13 "looks like." However you do it, I urge you to keep a log of some kind. Looking over my entries of 10 and 20 years ago is always a pleasure, and often I'm surprised and gratified by my past accomplishments: "I saw that from the city? With only a 4-inch reflector?"

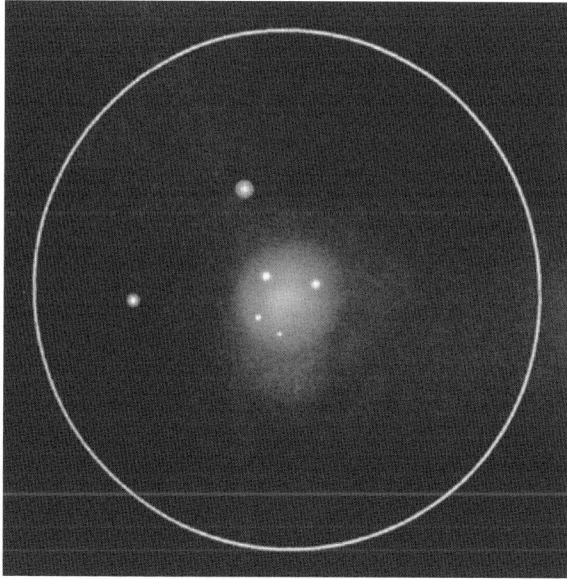

Figure 5.2. M10 drawing.

Astrophotography and Imaging in the City: The Ultimate Observing Project

Eventually, just about every serious deep sky observer dreams of making pictures of what he sees. Sure, you have fantastic memories, log entries and maybe even drawings of the hundreds or thousands of objects you've observed. But for some of us that suddenly doesn't seem to be enough. We want a *real* picture of M42 or M13. Your fellow visual observers may try to reason with you, warning you of the difficulties of astrophotography—especially in the city. But that won't matter. When the astrophotography bug bites, you won't care how hard picture taking is, and you won't care that you can get on the Internet and download as many pictures of M42 as you want. You will want your own pictures of Orion.

Actually, your seasoned amateur astronomer friends probably won't just tell you that picture taking in the city is difficult. They'll tell you that photography, film photography, is *impossible* from light-polluted areas. Impossible? Hardly. Difficult? Yes. You will have to accept the fact that, while you'll be able to make some wonderful pictures, they won't ever be as good as what you can get in the country. You'll also have to content yourself with capturing the brighter objects if you intend to use film.

Film Astrophotography

The first thing you need to get started in astrophotography is not a camera, it's a book. There are numerous good books on this challenging art available, as well as a lot of informative web pages on the Internet, but the best place to begin is with Michael Covington's classic work, *Astrophotography for the Amateur*. This book will tell you everything you need to know to get you going. I can't possibly adequately cover the complex techniques of successful celestial picture taking in just a few pages, but, if you've schooled yourself in the basics of the practice with Covington's book or other sources, I *can* tell you how to go about deep sky photography *in the city*.

In the city, it all comes down to the *speed of your telescope*—it's focal ratio. As we know, a scope with a "slow," large focal ratio produces bigger, more magnified images with a given eyepiece than a fast, small focal ratio telescope. In fact, a large f/ratio scope produces bigger images even *without* an eyepiece, during prime focus photography when the camera's film takes the place of the eyepiece. A small focal ratio scope projects a smaller, brighter image of a nebula or galaxy onto the film. Due to the smaller, brighter image, the picture builds up more quickly, *faster* in a "fast" small focal ratio scope than in a telescope with a larger focal ratio, one that projects *larger*, dimmer objects on the film. A large image builds up *slowly* on film. A large focal ratio scope is a "slow" scope. Normally, deep sky photographers want to use a fast telescope, one with a focal ratio of $f/8$ or $f/6$ or even faster. Not in the city. Using a relatively slow scope, one with a focal ratio larger than $f/8$, maybe one as slow as $f/10$, is the key to getting decent pictures.

The problem in the city for astrophotographers, just as it is for visual observers, is sky glow. In a fast telescope, an image builds up in a hurry. But so does evidence of light pollution. This comes in the form of "sky fog," as seen in Plate 22. I took this image, a 10 minutes exposure, through an 80-mm $f/5$ refractor from the heavily light-polluted site shown in Plate 1. Sure, you can make out the Orion Nebula and the stars of Orion's sword, but the background is bright and milky in appearance and obscures even medium bright stars.

What can you do? One approach is to reduce exposure time. That can help, but it's not the answer. The problem with shorter exposures is that it becomes very difficult to capture dimmer objects. The Orion Nebula is incredibly bright at an integrated magnitude of 4.0, so even a short exposure with a high-speed scope like an $f/5$ allows a recognizable image to build up on the film. Most DSOs are not nearly so bright, and exposures short enough to reduce the film-fogging effects of light pollution will not adequately record the object of interest.

Another problem with fast telescopes, especially those with small apertures like the 80-mm, is that they produce small images. That is OK for large galaxies like M31 or emission nebulae like M42, but becomes a problem when you're trying to record small galaxies and planetary nebulae. You can enlarge the image in the darkroom or computer, but there won't be much detail visible no matter how much you blow up the negative. The image scale was too small for good resolution.

Is there any way to image dimmer objects in the city while preserving image scale and preventing sky fog due to light pollution? Yes. The way to achieve good-looking deep sky photos in the city is to image with a slow, large focal ratio scope. How does a larger focal ratio with its longer focal length help in the city? For one thing, it *spreads out* the sky glow, reducing its intensity, allowing much longer exposures before sky fog becomes apparent. This is similar to our technique of reducing the intensity of the bright sky background visually by using higher magnification eyepieces.

Plate 23 is a 15-minnute shot at $f/10$ (with an 8-inch Schmidt Cassegrain). While there is some obvious fogging from sky glow, I don't feel it hurts the picture much, and what little there is could be easily removed by scanning the picture into a computer and "processing" the image with a program like *Adobe Photoshop*. Another benefit of larger focal ratios is that even big objects like the Andromeda Galaxy show off more detail when you take advantage of all the resolution capabilities of your film by enlarging your pictures in the camera using more focal length.

What's the maximum length of time you can expose an image in light pollution before fogging becomes unbearable? This depends on the focal ratio of the scope, the speed of the film (its ISO/ASA number), and the condition of the sky. The larger the focal ratio of the telescope, the longer you can go. So why not go to go an $f/15$ or larger focal ratio? Because any telescope eventually reaches a point of diminishing returns.

The larger the focal ratio and the longer the scope's focal length, the bigger and dimmer the image at prime focus gets. With an $f/15$ scope you can expose considerably longer than 15 minutes, even from the worst urban settings. Unfortunately, the images become so large in scale that it will be difficult to properly frame anything other than the smallest objects. The Orion Nebula will spill off the negative's edges. Why take a picture of M42 if you can only get the innermost core of the nebula into the shot?

More importantly, the larger the focal ratio and longer the resulting focal length, the longer it takes to build up an acceptable image on the negative. At $f/10$, it's possible to go as long as 30 minutes in medium-heavy light pollution. Even 30 minutes, though, is not long enough to adequately record the dimmest Messiers at this focal ratio. They become too big and too dim at $f/10$. Go to really slow scope, an $f/15$, with its even bigger and dimmer images, and only the brightest sky objects can be properly recorded with reasonable length exposures. Sure, exposures of 1 hour may be possible at $f/15$ and higher focal ratios, but, even after an hour, most objects will be dim and disappointing on the negatives. The DSO is too big and spread out to look good, even after a marathon exposure.

Another variable is film. The speed of the film will affect the length of your exposures. I prefer a medium-speed emulsion in the ISO 200 range with good red sensitivity for nebulae. A 200 speed film is fast enough for most deep sky work, and it helps keep the sky fog down. Go to ISO 800, even with an $f/10$ telescope, and your urban shots will fog up in a hurry. Unfortunately, there are few films for the astrophotographer to choose from lately. In the last few years, the major film companies, Fuji and Kodak, have changed the emulsions of many of their films to reduce their red response. This may be good for terrestrial photographers, but it makes a film almost unusable for us. How can you record the beautiful reds of M42 if your film doesn't respond to that part of the spectrum?

The best 35-mm film available today for the astro-imager is Kodak's Elite Chrome 200 transparency ("slide") film. Unfortunately, my tests with it show that the film's spectral response in the red is relatively poor, and that it seems to pick up sky glow more readily than some other films I've used. Compare Plate 23 (Elite Chrome) to Plate 24, which I made of M42 6 years ago from a similarly light-polluted site. The image in Plate 24 was made with the same telescope, camera, and exposure time (20 minutes) as Plate 23, but the film used for Plate 24, Fuji's Super G800, recorded more nebulosity and noticeably less sky fog. Unfortunately, the Fuji film is no longer produced. Kodak and Fuji are both switching over to digital imaging products, and the roster of available films is shrinking every day. The emulsion types useable for deep sky photography are disappearing even more quickly. Elite Chrome is currently the best of a disappointing lot.

Are there any other ways to improve urban images other than increasing scope focal ratios? A few urban astrophotographers report good results from exposing through light-pollution reduction filters. I've tried these filters with cameras, but was disappointed with the results. The most serious limitation in using these filters for photography is the "filter factor." Introducing even a mild filter like the Lumicon Deep Sky or Orion Skyglow into the light path means the light from the target object is badly dimmed before it reaches the film. As with high focal ratio scopes, long exposures are possible through filters, but it will *take* much longer for a suitable image to build up.

Another problem in using LPR filters for picture taking is color shift. Most deep sky astrophotographers are now forced to use color film, since the single black and white emulsion suitable for deep space imaging, Kodak's Tech Pan, has, like so many other films, been discontinued. Use an LPR filter with color film and images will assume a very strong red or green cast depending on the particular filter in use. This "excess color" can be removed by computer processing, but a light-pollution filter often leaves the negative with such a strong color bias that it's difficult to bring it back to "normal."

How do you determine the required exposure for your site, scope, object, and film? The best bet is experimentation. Expose a DSO in 5 min increments. Say 5, 10, 15, and on up to 20 or 30 minutes (at $f/10$, less if you're shooting at faster, smaller focal ratios). Examine the results and decide which exposure gives a good balance between sky fog and detail. One final bit of advice? Make sure the telescope is well shielded from ambient light. Light from nearby sources leaking into your telescopes will make even the shortest exposures with the slowest scopes look horrible.

Piggyback Astrophotography

Do you have to photograph *through* a telescope? If the above sounds too challenging, you can start out with simple piggyback imaging. In piggyback astrophotography, you mount a camera on the scope's tube via an inexpensive bracket and it photographs through its own lens rather than through the telescope. This provides a good introduction to astrophotography without the difficulties associated with precise polar alignment and guiding, and most deep sky imagers have traditionally started out this way. Piggybacking is not only easy, it's a good way to take attractive, wide-angle constellation shots.

If you intend to try piggyback astrophotography, be aware that the fast focal ratios of your camera's "normal" lens— $f/2.0$ or less—mean that sky fog will build up in a tremendous hurry. Go much more than 30 sec of exposure time, and you'll swear the resulting prints were shot in the daytime. Plate 25's shot of Comet Hale Bopp was taken well after dark despite its apparently blue skies. If you want to try piggyback imaging in the city, use a slower telephoto lens. You may find old, "slow" longer focal length lenses with focal ratios of $f/5$ to $f/8$ available for very reasonable prices in your local camera store's used department.

The CCD Alternative

You can get some attractive film images in the city, as I hope my Great Nebula pictures show, but there is no denying that the film astrophotographer is limited by light pollution. Due to sky fog, you must keep your exposures short, even with slow scopes. Forget imaging beautiful but dim objects like the Horsehead Nebula, or even capturing good detail in brighter objects. As many astrophotographers are finding, though, there is a substitute for dark skies when it comes to picture-taking, the CCD camera. The CCD camera and its use, like astrophotography in general, is a subject for an entire book. I do, however, have enough space and experience to talk briefly about why these electronic cameras are of extreme interest to urban astronomers.

The CCD camera is a camera that uses an electronic chip, a "Charge Coupled Device," instead of film. This CCD chip is similar or identical to the one in your video camcorder or your "digital" still camera. The only major difference between an astronomical CCD camera and your digital snap-shooter, in fact, is that an astronomical CCD camera is usually equipped with a device that cools the chip, often to well below zero. One other difference is that most astronomical CCD chips deliver black and white pictures rather than the color images of camcorders and digicams. This is because black and white CCD chips offer better resolution and sensitivity for astronomical subjects.

Why cool a CCD camera? Heat means noise when it comes to digital images, and the longer you expose, the worse it becomes. After just a few seconds, the image produced by an uncooled camera is covered with "snow"—speckles representing thermal noise. For the terrestrial still camera, this is not a problem. You rarely need to exceed a few seconds of exposure time due to the sensitivity of CCD chips. Some of these chips have equivalent ISO/ASA ratings in the thousands. But even with this sensitivity, an astronomer may need to expose a faint nebula or galaxy for 5, 10, 15 minutes or longer. At these exposure times, an uncooled camera's image would be buried by thermal noise. So, long exposure astronomy cameras have a means for cooling them down to defeat thermal noise. This cooling may be provided by water or air, but is most often accomplished by a device called a Peltier chip. The Peltier is a thermoelectric cooler that can chill a CCD to well below freezing without the need for liquid coolants, eliminating most heat-produced noise.

You've got a very sensitive camera capable of exposing for as long as you want to go. But how does this help the urban imager? What makes CCD better than film in the city? The most important benefit for the urban astrophotographer is the chip's sensitivity. You can keep exposures short, even with a relatively slow telescope, but still pick up plenty of details before you see evidence of background sky fog. Another help is the relatively small size of CCD chips. This increases the image scale, just as if you increased your scope's focal length, further suppressing light pollution's effect. Finally, CCD cameras produce digital images with very good *dynamic range*. What that means is that any problem with light pollution can be easily reduced or totally eliminated by standard computer image processing programs. The image in Plate 26 was taken from a light-polluted site even worse than the one I used for the earlier film images, yet shows little or no sky fog.

CCD cameras are not magic, of course. Their biggest drawback is their expense. An entry-level cooled camera with a small chip (about 1/5 the size of a 35-mm film frame) goes for around 1,000 US$. Cooled astronomical cameras will always have a limited market so don't expect prices for entry-level cameras to go down much, though chip-sizes and features at a given price-level will continue to increase. You'll also need to take a laptop computer into the field, as almost all astronomical cameras require a computer to control the camera as well as save and process images. Finally, for those of us brought up with the beauty of film astrophotography, even the most expensive CCD cameras in the most experienced hands have yet to produce pictures with *quite* the beauty and depth of traditional emulsion astrophotography. On the other hand, a CCD user with a medium-sized scope can rather easily record stars of 18–20th magnitude, something that was difficult for professional observatories under the darkest skies and equipped with giant telescopes 30 years ago.

Other Electronic Imaging Alternatives

What if you want to take non-film images in the city and don't think a traditional CCD camera is for you? As of now, there are three methods of image making that improve on photography but which don't demand the financial investment required by "real" integrating astronomical CCD cameras.

Both amateur and professional astronomers have used video to take pictures of the sky for the last 30 years. In the mid-1990s, I was one of a group of amateur astronomers who demonstrated that the newer video camcorders with their relatively sensitive color CCD chips and high definition tape formats (Hi 8 and Super VHS at the time), could take images of the Moon and planets that easily exceeded anything that could be done with film, no matter how large the telescope. While this was a breakthrough for Solar System imagers, it wasn't much help for the deep sky photographer.

Shooting at 1/30 second in color, the family camcorder is unable to produce a good image of even the brightest DSOs. Some amateurs experimented with low-light surveillance cameras, black and white "closed circuit" cameras that are much more sensitive than the average camcorder, and had some success. They would shoot many minutes of footage, download this video to a computer, and combine hundreds of frames into acceptable deep sky images. This was encouraging for video fans, and seemed to work as well in the city as under dark skies, but the pictures produced, though nice, were a long way from the images coming out of the cooled CCD cameras of the time.

Then everything changed. Sony began producing a very sensitive CCD chip intended for use in video cameras, the ICX248AL. This CCD is at least twice as sensitive to light as a conventional chip, including those used in astronomical cameras. The Sony CCD is also surprisingly inexpensive, so it wasn't long before it was being used by several manufacturers in cameras designed for astronomy, notably Adirondack Video Astronomy in the U.S. with their Stellacam and Mintron in the UK.

What makes these cameras innovative is not just the fact that they are incredibly sensitive, it's that they go well beyond what a normal video camera can do. In addition to shooting video sequences, they are able to automatically add frames together internally and thus deliver the equivalent of long exposures. Things have gotten even better recently, with the latest "deep sky video cameras" featuring even more sensitive chips and longer exposure capability.

The pictures produced by these cameras are surprisingly noise free considering the fact that they are uncooled, and are definitely comparable to those taken with entry-level cooled cameras (see Plate 27). They actually have several advantages over conventional CCD setups. Each exposure taken by a Stellacam or Mintron is short, so fewer demands are placed on a telescope mount's ability to track precisely. You don't need a computer, either. While they can be used in conjunction with a computer (equipped with frame-grabber hardware), the images they produce can also be recorded with a simple home VCR. Some users don't even worry about recording the images. They use these cameras for real-time deep sky observing.

Real-time deep sky viewing with a video camera may not be quite as "romantic" as old-fashioned visual observing. On the other hand, the view of a dim object like galaxy M51 on the video monitor in the city is much more detailed than you'll see through an eyepiece under the darkest skies. The spiral arms of this object will be far more clearly seen than they will visually in a scope 3 or 4 times as large as the one the Stellacam or Mintron is imaging through. Since these cameras output large-format real-time video, focusing and framing is much easier than with an integrating CCD camera. At about half the price of a conventional astronomical CCD imager, the Stellacam is well worth a look by the urban imager.

Everybody I know has got a digital still camera these days. The filmless camera is all the rage, and prices are declining while quality and features are increasing at an amazing pace. Naturally, the first thing an amateur astronomer thinks of when getting her hands on one of these new megapixel wonders is, "Can I take astrophotos with it?" Initially, the answer was, "Yes, but only of the Moon and planets." Digicams are uncooled, and the initial products aimed at amateur photographers were not able to take exposures longer than a second or two.

Astrophotographers have never been stopped by conventional wisdom, however, and it wasn't long before they were taking amazingly beautiful *deep sky* pictures—color deep sky pictures—with digicams. How? Using the same technique video imagers had used before them: taking many shots and combining—"stacking"—these frames with PC imaging software like *Adobe Photoshop*. The deep sky pictures coming out of digital cameras just keep getting better. Not only are astrophotographers developing innovative ways around the thermal noise problem, top of the line amateur-grade digicams now offer longer exposures (5 minutes or longer) and built-in noise reduction software. Just like "real" CCD cameras, digicams are relatively immune to skyglow problems.

That said, digital camera still cannot produce images as good as those of dedicated astronomical CCD cameras. They do very well on bright objects, less well on dimmer ones, and noise is still a problem. The average digicam image requires far more processing than that from an integrating astronomical camera. The digital cameras best suited for astronomy work, the digital Single Lens Reflexes, are still fairly expensive, too, though their prices are beginning to come down now. If you're interested in a camera that can take good-looking color images of showpiece DSOs, and also do a wonderful job for terrestrial picture taking, one of the newer digital SLRs might be right for you. I would not, at this time, buy one of these cameras just for astronomy, however.

Don't have a digicam? Don't want to buy a Stellacam? Sure don't want an expensive cooled camera? Still want to take deep sky pictures? A webcam, one of those little video cameras designed for use with PCs for teleconferencing can bring back amazingly good portraits of DSOs. When these devices first became widely available 10 years ago, curious astrophotographers decided to see if they could be used in astronomy. They had one major advantage over video cameras—they are designed to output their images directly into the computer in digital form. No videotape, frame grabber, or other intermediary is required. Initially, as with video, it appeared that the webcam would be useful only for the Moon and planets, due to its inability to shoot video at exposures much longer than about 1/5 second.

Again, amateurs found workarounds. One thing that works is to do the same thing video and digicam users do—stack images. That produces surprisingly nice results. But the webcam crew was not content to stop there. Creative amateurs have found ways to electronically modify webcams to allow longer exposures, and have even added Peltier coolers to them to cut down on thermal noise. Excellent deep sky images are being done in the city—in color—with webcams purchased used for as little as 10 US$.

If you'd like to learn more about webcams in astronomy, the people to talk to are the members of the Internet QCUIAG, the "Quick Cam and Unconventional Imaging Astronomy Group." They can help you select the right webcam. Some are better than others, and many current webcams use CMOS imaging chips rather than the CCD chips that are best for astrophotography. Webcams have gone from producing images that impressed just because they were identifiable pictures of familiar DSOs, to images that can stand up against those produced by much more expensive cooled cameras.

Recently, we've begun to see cameras that blur the boundaries between the integrating astronomical camera and the webcam. The most exciting of these new cameras, the Meade Deep Sky Imager, is about three times the cost of an off-the-shelf webcam, but is capable of long exposures out of the box—no modifications required. The DSI also features a larger chip than most webcams, though its CCD sensor is still smaller than that of most integrating cameras. The DSI normally produces color images, though it can be told to output monochrome instead.

The software included with the DSI is what makes it special. Its innovative "suite" of programs allows it to operate in a manner that combines the best features of the webcam and the integrating camera. It can output near real-time video, making it easy to find and focus objects, but it can also do long exposures of up to 1 hour in duration. It can also guide your telescope if you're taking images with another camera, or guide the telescope between its own exposures, making corrections when it's not taking pictures. The DSI is not actively cooled, but it does have an innovative passive cooling system that appears to drastically reduce thermal noise. The Meade Deep Sky Imager appears to be a big hit with amateurs, so expect to see similar innovative and inexpensive cameras from other manufacturers.

"Five chapters of talk! When do we get to the *good stuff*?" I hope the first part of this book has been helpful regarding the techniques and equipment needed for rewarding urban observing. As you've seen, it's not as simple as just grabbing the telescope and running for the backyard. But now it's time for that good stuff. What comes next is a "walking tour of the cosmos from your bright backyard," a feast of wonders for every season of the year. In the beginning I told you you'd be surprised at what can be seen from the light-polluted city. But "surprised" is not a strong enough word. Prepare to be *amazed*.

Before getting started on the Tours section, you may want to have a look at Appendix 2, which explains directions and distances in the sky, something that is very important for object locating if you're star hopping. You may also find Appendix 3 helpful, as it explains the classifications and "codes" used to describe DSOs in this book and in other resources.

Which telescopes did I use to observe the objects in the Tours part of this book? The City Lights Telescopes (Plate 28) were a 4.25-inch f/11 Newtonian reflector, a 6-inch f/8 Newtonian, an 8-inch f/5 Newtonian, a 12.5-inch f/4.8 Newtonian, an 8-inch SCT, an 11-inch SCT, a 60-mm refractor, and an 80-mm refractor. Which ones got the most use? The 4.25-inch reflector and the 11-inch SCT. I used the small Newtonian whenever possible to try to gauge how small an instrument would reveal chosen targets. The 11-inch Celestron Schmidt Cassegrain brought sky glow defeating aperture and a computerized go-to drive system to my bright sky survey, and conquered the most aggravating faint fuzzies.

What exactly will you find in these tours? All types of DSOs and more. Should double stars be classified as DSOs? I don't know, but they sure are beautiful, so I've included quite a few for your enjoyment. The main course here, though, is those astoundingly beautiful galaxies, nebulae, and star clusters you crave, well over a hundred of them. You'll find a complete list of the objects visited in this book in Appendix 4 along with their vital statistics including their Right Ascensions, declinations, and magnitudes.

Grab those eyepieces, would you? No, you won't need a coat—it's almost always shirtsleeve observing weather down here in sunny southern Alabama. It's time to walk the deep sky.

PART II

A Walking Tour of the Cosmos

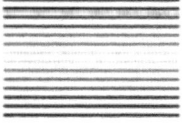

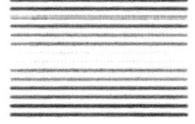

CHAPTER SIX

Spring

Tour 1

Burning Heart of the Hunting Dogs, M94, M51, and Company

In the city, the advent of the spring observing season is both a delight and a disappointment for the amateur astronomer. A delight because this time of year brings warm weather, making observing a much more pleasant experience. Spring also delivers "new" constellations and signals the return of the great forest of galaxies that stretches from the northernmost Ursa Major to southernmost Hydra. A disappointment because the incredible, delicate riches of the spring deep sky are badly dimmed by the senseless light pollution of our cities.

If you like galaxies, this is an especially frustrating time. They suffer tremendously from light pollution, so badly that it sometimes seems easier to stay inside and watch television, avoiding mosquitoes and aggravation. As we'll see in this chapter, however, there are many lovely galaxies all through Coma and Virgo available for city-bound sky watchers equipped with medium or even small aperture scopes. But before launching ourselves into the awesome Virgo–Coma Cluster, let's travel the northern edge of the spring sky, the Canes Venatici and Ursa Major area. Both constellations offer bright galaxies aplenty, but with a couple of bonuses: a good planetary nebula and a globular star cluster are also on display. Once you get into Virgo, it's almost nothing but galaxies.

M94

Don't think your small scope is capable of showing you galaxies from your bright backyard? Think again. M94 is different. This galaxy has always been one of my first stops when the handle of the Dipper/Plough ascends above the horizon haze and I can again see the two prominent stars that make up Canes Venatici, the Hunting Dogs. Why do I like M94 so much? This sucker is *bright*. Magnitude 8.9 packed into a size of 5′ × 3.5′ means it stands out well, no matter how bright my skies are. I recall at least one time when searching for this galaxy with my first scope, a 3-inch reflector, that I almost gave up. Instead of the dim blob I expected, M94 looked like a prominent star and I kept passing over it before realizing my error.

M94 is surprisingly easy to find. The only prerequisite for locating it is that you must be able to make out Canes Venatici's rather unspectacular pattern of stars. Referring to your star atlas or the chart in Figure 6.1, look southwest of the Dipper/Plough's handle to Canes Venatici's two lone bright suns, Alpha Canum Venaticorum (Cor Caroli) and Beta Canum Venaticorum (Chara). Cor Caroli, by the way, is one of the finest double stars in the Northern Hemisphere. A pair of bright sapphires (magnitudes 2.89 and 6.0) separated by almost 20″, this binary is a treat in nearly any telescope.

Referencing your chart, note the position of M94 with respect to the constellation's two bright stars, and you'll see that it lies approximately a degree and a half outside the line connecting Alpha and Beta, and is back in the direction of the handle of the Dipper, northeast of the line. Before going to the scope, make sure you have your

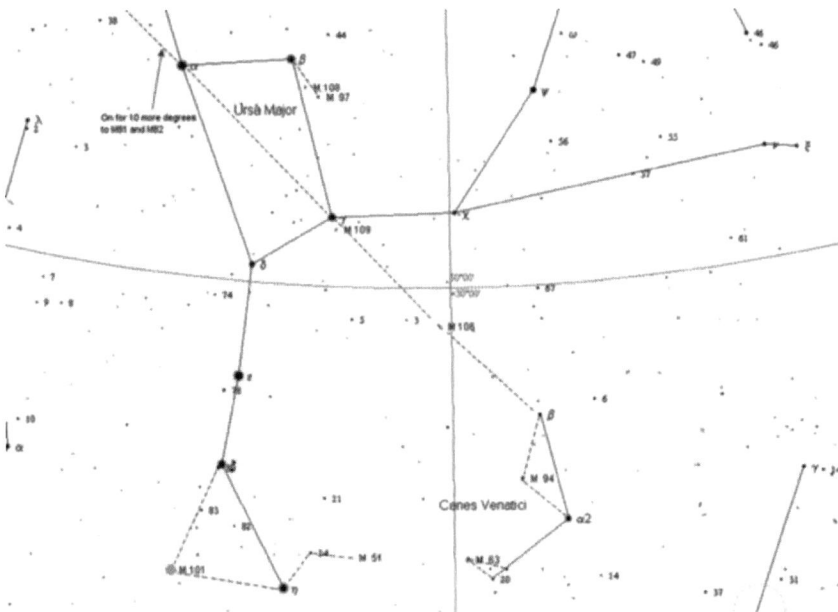

Figure 6.1. *Canes Venatici and Ursa Major.*

chart oriented correctly, inverted for most finders. Try to fix M94's location against the stars firmly in your mind.

Once you're properly positioned, take a look through the main telescope, starting with an eyepiece that yields a magnification of approximately 50×. What you should see is a nebulous "star" in the field. You may have to sweep the scope around a bit, but if you've positioned it carefully using finder and atlas, it shouldn't take much looking. This star-like object is M94. Let me repeat myself, M94 will look bright, *much* brighter than you'll probably expect a galaxy to be. If you have a hard time deciding whether what's in your field is really the galaxy or just an anonymous star, increase your magnification to 100–125×. That will make its true nature pretty obvious. Averted vision should show a fairly large nebulous envelope around the galaxy's stellar-appearing core.

In my 4.25-inch Newtonian reflector, M94 looked amazingly like a lot of smaller globular star clusters do in small scopes under city lights: bright, round, and surrounded by an easy-to-see mist of "unresolved stars" (really unresolved in this case!).

My log entry for M94 with the 4.25-inch Newtonian reads:

> Almost stellar at low power. At 100× it is impressive in this small aperture. Brightens smoothly toward a brilliant, almost star like, center.

It looked even nicer in the 11-inch Schmidt Cassegrain:

> In the C11, M94 is amazingly, bizarrely bright. Its preternaturally bright nucleus just blazes away. Also visible is a fairly extensive outer envelope of nebulosity. I strain for some hints of its spiral arms, but no other details are visible in this perfectly round galaxy, even at 300×. It is still spectacular for its brightness compared to the way most other galaxies look in the city.

M94, a Hubble Sb type spiral, is thought to be approximately 14 million light-years distant. In long exposure photos taken with professional telescopes, the bright disk is surrounded by tightly wound spiral arms. These arms are not overly easy for the amateur astrophotographer and are nearly impossible visually in the city, even with large apertures. This is due both to their constricted nature and the huge contrast difference between them and the galaxy's core.

Why is this galaxy's nucleus so bright? It is believed that the center of M94 harbors a monstrous black hole that is currently "feeding" on gas found there or perhaps even gobbling stars unfortunate enough to fall within its gravitational grip. While our views of the central portion of our own galaxy are spectacular, those from a planet orbiting one of M94's stars must be indescribable. Their "Sagittarius Region" would blaze forth, with the fires of nuclear annihilation, though most of the glory there would be obscured by dust clouds, just as the seemingly sedate core of our own spiral is hidden.

M51

As a boy, the galaxy I *most* wanted to see was the Whirlpool, M51. I spent a lot of time staring at a photograph of this object taken with the 48-inch Oschin Schmidt Camera at Mount Palomar (Plate 29) and dreaming of what the Whirlpool Galaxy and its

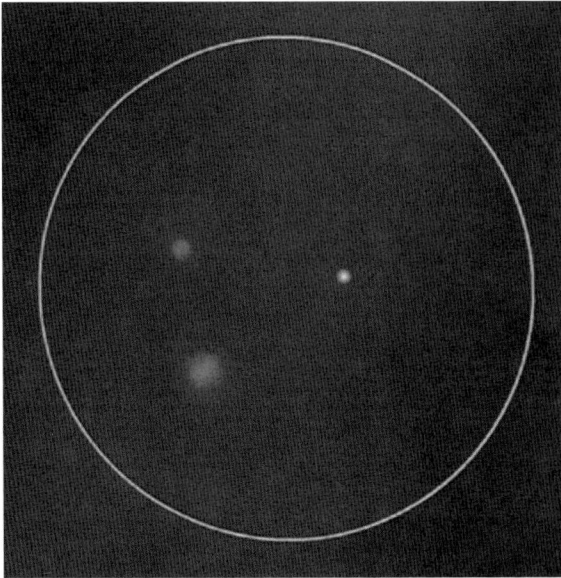

Figure 6.2. M51 with pencil and paper.

small companion, NGC 5195, would look like in my own telescope. Those majestic spiral arms really fired my imagination. Sadly, I was never able to see it at all, much less detect its spiral arms, with my 3-inch or (later) 4.25-inch telescopes. Why? I didn't know *how* to look for it or what it would look like when I found it. In the relatively good, though still light-polluted, skies of my youth this object would, admittedly, not have been overly easy in such small instruments, but would have at least been visible if I'd known how it appears in little telescopes.

M51 looks marvelous in the photos, showing off both its magnificent spiral arms and a bridge of material stretching between it and the passing irregular galaxy, NGC 5195. From a dark observing site it's wonderful visually, too, with the spiral shape being apparent to an experienced observer equipped with a scope as small as 6 inches. Sadly, I have to admit that in the city it is something of a dud. There are two strikes against it. It's large at 11′ × 7′ minutes in extent, and it's a face-on Sc galaxy, which means its supposedly bright (integrated) magnitude 8.9 light is badly spread-out and diluted.

Does this mean it isn't visible from the average urban site? No. I've seen it with a 6-inch Newtonian reflector from a location less than one mile from a brilliantly illuminated shopping mall (and with a 4-inch scope from a slightly better site). I just don't want you to imagine that M51 will look *anything* like its pictures—or even like its visual appearance from dark sites—in your urban telescope. You will also have to *work* for it, but you can do it, and will at least be able to see the central regions of both M51 and 5195 (Figure 6.2).

If it were a little brighter, M51 would be relatively easy to locate, as bright Eta Ursae Majoris and plenty of other prominent guide stars are visible in your finder in this area to lead you to it. Since the galaxy is surprisingly dim, though, you must take extra care when positioning your scope. M51 is 3°37' to the southwest of Eta Ursae Majoris, but a better guide to it is magnitude 4.7 24 Canum Venaticorum, 2°9' almost directly west of Eta. Hop to 24 from Eta, and, still looking through your finder, you'll see a pair of 7th magnitude stars 47' apart. M51 lies slightly outside a line drawn between these two, and closer to the easternmost star. Position your scope, and insert a medium-power eyepiece into the focuser. If you're lucky, you should see a painfully dim double glow in the field. That may be all you'll see in smaller scopes, as the brightest star in M51's telescopic field is at 12th magnitude. Nothing? Recheck your positioning with regard to the two 7th magnitude stars, and, importantly, increase your magnification. I frequently found it took at least 100× to reveal M51.

What do you get for all this work?

> This is a very good night, and M51is riding high in the sky. It's easy to see in the 6-inch f/8 Newtonian. For once, I didn't have to guess at to whether I was really seeing it or not! Two round spots of nebulosity, one larger than the other, with no core noted for either object.

An 11-inch SCT makes the galaxies brighter if not much better:

> At 127× in the C11 with a 22-mm Panoptic eyepiece, M51 and its companion look much the same as they do in a smaller scope: a big fuzzball and a little fuzzball. The difference is that they are much easier to see and find in this telescope, and that averted vision brings out quite a bit of barely-perceptible nebulosity beyond the main galaxy's core that's invisible in smaller apertures. M51 also occasionally displays a star-like nucleus, something I don't recall ever seeing in a 6-inch scope from town.

M51, a Hubble Sc class spiral, is floating 35 million light-years out in the darkness. Its accompanying galaxy, little NGC 5195, has, as is obvious from the "bridge" of material connecting the two, recently (in the last few million years, that is) had a close encounter with the larger galaxy. Despite appearances, NGC 5195 does not lie in the same plane as M51—it is in the background and is receding.

M106

M106, is not nearly as famous with deep sky observers as M51, but it is very interesting in its own right. Never looked at it? That's a shame, since it's really much more attractive than M51 in the city. It's easier to find as well. Located in the northwestern part of Canes Venatici, this Sb spiral is large but bright at 17.4' × 6.6' and magnitude 8.3. A glance at a chart locates M106 near the center of a line drawn between Gamma Ursae Majoris (one of the Dipper asterism's bowl stars) and Beta Canum Venaticorum. Unlike M51, M106's surface brightness is remarkably high, so you may be able to use a searching power as low as 50× to make the hunt easy.

How does M106 look in small telescopes? Great! For quite a few years I skipped over this cosmic beastie. I didn't hear fellow amateurs mention it often, and it isn't even one of the original Messier objects (numbers 104 and below), so why bother? One evening, after struggling fruitlessly with M51 with the 6-inch scope from a very poor site, I remembered nearby M106. When I had the galaxy in the field of a 12-mm Plossl, I was very sorry I'd ignored it for so long:

> M106 really is quite bright and is easy to find and identify. Plainly visible with direct vision even with tonight's poor skies. More or less round with some hint of elongation. Compact core visible.

And all this with the lights of my city's shopping mall providing a backdrop. If your scope or eyes or observing site are better than mine, you may be able to see another, smaller galaxy, NGC 4248, 14′ to the west and in the same medium-power field with M106. Unfortunately, at magnitude 12.5, NGC 4248 will be tough (I won't say *impossible*) from most urban locations. There are many, many galaxies scattered across this area of the sky, but most of them will remain invisible in the streetlight glow, I'm afraid.

M63

M63 is another galaxy that, while at its best from dark sites, bears up remarkably well from a bright backyard. This is the famous "Sunflower," a name it bears because of the curious appearance of its spiral arms. They are fairly tightly wound and have a splotchy, clumpy appearance. Combine this with the bright inner disk of the galaxy, and, yes, it looks a little like a great cosmic sunflower. While the flower-aspect of M63 is mostly reserved for darker skies, I find myself catching hints of it from time to time with the 11- or 12.5-inch telescopes from the city.

Finding M63 is not a huge challenge. It's $1°29'$ from magnitude 4.73 20 Canum Venaticorum, and forms a slightly lopsided triangle with 20 and bright Cor Caroli, which is $5°17'$ to the south-southwest. This large (12.6′ × 7.5′), relatively bright (magnitude 8.6) nearly face-on Sb galaxy is apparently a member of the small group of galaxies that includes M51. For best results, view it at about 125×, though it's bright enough to be found at lower powers. Keep coming back to this one, especially on above average nights, and you'll eventually be rewarded with a clear look at its spectacular sunflower aspect. On average nights, your looks at it will probably be more like mine on a typically muggy May's eve':

> From my heavily light-polluted backyard, M63 doesn't reveal any sign of its spiral arms in the 12.5-inch Dobsonian, not at first. The disk is fairly large, about 5′ across, and I can see both a smooth brightening toward its center and the occasional hint of a tiny point-like nucleus. The galaxy is obviously and strongly elongated east/west. As I stare at the Sunflower, using a range of eyepieces from 80× to over 200×, I actually do begin to think I'm seeing hints of spiral structure, but it is incredibly subtle, and may have more to do with what I remember from photos and from observing this object from dark sites than with what I'm actually seeing in the eyepiece tonight.

M81

Heading out of Canes Venatici and over into Ursa Major, we find M81, one of the most beautiful galaxies in the northern sky when seen from a dark—really dark—site. It has a bright core, a large oval outer envelope of nebulosity, and two gossamer spiral arms. Even from pristine desert skies, M81's arms are subtle, and they are, unsurprisingly, utterly invisible from the city. This is still a nice target, though, and simple to find despite the fact that it's in the seldom-visited northwestern part of the Great Bear, well away from the Dipper/Plough asterism.

To find M81, draw a diagonal line through Gamma Ursae Majoris, Phad, the southwestern bowl star of the dipper (or "blade," if you see a plough here instead) and through Alpha, Dubhe, the star diagonally opposite Phad, which is 10° from Phad. Extend this line *another* 10° into empty (mostly) space and you'll land in the area of M81. You may have to search around for a while, as the nearest prominent star to the galaxy is magnitude 4.56 24 Ursae Majoris, fully 2° to the northwest. Luckily, M81 usually stands out well enough that slow, careful slewing-around will pick it up without too much difficulty. A medium-power eyepiece should reveal M81's softly luminous 24.9′ × 11.4′ disk (that's its size in photos, anyway) glowing cheerfully at magnitude 6.9.

In the field of my C11, M81 on a below average night was still interesting:

> M81 is beginning to sink in the northwest, and is now into some of the worst of the light pollution. Beautiful, but the outer envelope of nebulosity is dramatically reduced in extent from what's visible from a dark site. No hint of spiral arms. Best view is in a 12-mm Nagler at 233×, which reveals a bright core at the center of this galaxy's misty, elongated disk.

While I used the C11 on this particular evening, I've seen M81 with remarkable ease on a good city night with my tiny 60-mm Meade ETX refractor.

M82

If M81 is wonderful from a dark site, M82 is mind-blowing. It is a magnitude 8.4 nearly edge-on galaxy of the Hubble Irregular type. This isn't your average edge-on spiral. Something *bad* has happened to M82. Its magnitude 8.4, 10.5′ × 5.1′ disk is criss-crossed by numerous dark lanes, which give it a "boiling," disturbed appearance. M82 is a strong radio source, indicating that it has indeed been violently disrupted. Years ago, this galaxy was thought to be "exploding," but today's astronomers believe M82's troubles are the result of an encounter with nearby M81 at some time in the distant past.

If you can find M81, you can find M82, as it's only 37′ away, just one medium-power eyepiece field to the northeast. If you've got a good night and the galaxies are at or near culmination, you may even be able to use a low-power eyepiece to see both of them at once in a wide field—a not to be missed treat. Like M81, M82 is routinely visible in small telescopes, as it was one early and chilly spring evening with my 4-inch scope:

> This must be an exceptionally good night, since I'm able to see both M81 and M82 without trouble, and can even put them both in the same field of my lowest power eyepiece, a 25-mm Kellner at 48×. M82 is mostly a featureless cigar, but by sticking with it and boosting my magnification with a Barlow lens, some of its dark lanes/patches do briefly come into view when I use averted vision.

M101

As a young observer I longed to see M101, the Pinwheel Galaxy, almost as much as I craved M51. If anything, M101's wide-flung spiral arms looked even lovelier in pictures than those of the Whirlpool. Unfortunately, just as with M51, I never even had a *hint* of M101 in the field of my 3- or 4-inch telescopes. This giant 22′ diameter face-on, Sc galaxy has a remarkably low surface brightness despite an integrated magnitude value of 7.9. It often doesn't look overly impressive from the darkest sites, even in 8-inch telescopes. Sky conditions must be *just right*, dark, steady and *dry*, for this one to strut its stuff. If you do have those things, though, it is an absolute marvel, filling the field of a medium-low power eyepiece and readily showing spiral arms in 8-inch and larger instruments. These incredible arms are peppered with obvious HII regions, huge patches of nebulosity analogous to our own galaxy's Orion nebula, which stand out dramatically when you view the galaxy through a nebula filter (usually a no–no for galaxies).

Here in the city, M101 is usually a "been there" DSO, one of those objects where the satisfaction comes from having been able to locate it, even if it doesn't look like much when you find it. To be able to brag to your friends that you've at least *been there*. As with all large Sc galaxies in the city, seeing it involves finding a compromise between an eyepiece with a *low enough* power to allow you to fit the galaxy in the field and provide some background sky to "compare" it to, and a *high enough* power to darken the background enough to provide contrast.

M101 is theoretically easy to pin down, as it forms a near equilateral triangle with bright Zeta and Eta Ursae Majoris, Mizar and Alkaid, the last two stars in the Dipper/Plough handle. Just position the scope to the northeast of these stars, as in Figure 6.1, and bingo! Right? I wish it were that simple. This is one you will probably *not* find on the first try—or tenth. It's too big and too dim. But everybody needs a challenge once in a while, so keep coming back, waiting for especially dark nights and for its culmination, and you will finally be rewarded with at least a glimpse of this majestic night-bird:

> Try as I might on this average night, I couldn't seem to find a trace of M101 with the 11-inch SCT. Not a trace. Finally, after using all the tricks: averted vision, dark hood, jiggling the scope, and switching eyepieces several times in a quest for the best magnification, I got a look at the central core. It's just a vague, nebulous ball about 10′ across, but I feel gratified to have defeated this monster again!

The Pinwheel Galaxy, was one of the last objects to be included in Charles Messier's original list of deep sky wonders, and was first seen by his friend Pierre Mechain in 1781. In addition to its scientific significance, M101 is historically interesting for a couple of reasons. It was one of the first of the spiral "nebulae" to be recognized as such. When William Parsons, the Earl of Rosse, the 19th century's most prominent amateur astronomer, turned his giant 72-inch reflector to M101 in 1845, he was able to see its spiral arms, and it became one of many of these curious objects that he recorded with his big telescope.

It is also famous because it is at the heart of a long running controversy over a "missing" Messier object, galaxy M102. There is nothing visible at the original coordinates given for M102 by Messier. While some astronomers, amateur and professional, have identified M102 with a galaxy in Draco, NGC 5866, it appears more likely that

M102 was actually a "reobservation" of M101 with the coordinates written down incorrectly.

M101 is one of a small group of galaxies lying about 24 million light years away from us. This big Sc spiral is approximately 150,000 light years in diameter, making it comparable in size to our own Milky Way. Think about that. Despite the glare of city lights, you've just visited the Milky Way's twin, seeing it as it appeared some 24 million years ago. Sure, it's just a smudge in your little scope, but, as we'll find time and time again as we hike the urban deep sky, your mind will fill in the blanks with wonder. After seeing it with your own eyes from your humble backyard, M101 becomes a real place, a majestic spiral clogged with suns. As you stare at its faint traces in the eyepiece, do you wonder if "someone" is staring back?

M97

Are you up for another challenge, albeit an easier one? Galaxies are great, but after spiral after spiral they all begin to look alike, especially in the city. If you want a break from island universes, Ursa Major harbors a bright DSO that's *not* a galaxy, the legendary Owl Nebula, M97. Surprised to hear the Owl referred to as bright? It really is, though it does have a reputation as one of the harder Messiers—an unwarranted reputation in my opinion. This Vorontsov-Velyaminov Type 3a (irregular disk with very irregular brightness distribution) planetary nebula is large—compared to most of its kind, anyway—at 3.4′ × 3.3′ in size, and you'd think this size, combined with its magnitude of 11.0, would make it tough. It is not. Not at all. In fact, I've seen it from the city with an Oxygen III filter-equipped 60-mm scope, the ETX 60, and it is routinely (though not always) visible without a filter in 8-inch and larger instruments in my messy skies.

You won't have much trouble getting the scope on the right spot for M97. It's 2°16′ south-southeast of an excellent signpost, magnitude 2.37 Merak (Beta Ursae Majoris) in Ursa Major's Bowl/Blade. Using your finder, position the scope on Merak and slowly move 2° due south. You should wind up with a fairly noticeable magnitude 6.63 star in the finder field. Put this star in the finder crosshairs, and, looking through the main scope's eyepiece, carefully move a half-degree (about one medium-power field, depending on your scope's focal length) south and east. When you're done, examine the field very carefully.

Don't see anything? Bump up the power a little and *really* look. With some trying you should see a large, dimly glowing disk staring back at you. If you've got a spectacular urban night, you may even see signs of the two dark spots in the disk, the "eyes," which give it its owl appearance and name. Let me emphasize the value of an OIII if you're after the eyes: without the filter I have never even *suspected* them in the city, not even in the 12.5-inch Dobsonian. With the filter, they're often obvious in that instrument and doable down to about 8 inches when conditions are right. The OIII can also make the difference between seeing and not seeing the nebula itself when conditions are poor or with smaller telescopes on any night. The nebula's central white dwarf star is at magnitude 16.0, so it's *probably* beyond our grasp in the city, even with fairly large apertures.

On this above average evening in the C11 SCT, the sometimes-frustrating Owl is a surprise. With an OIII filter and the 22-mm Panoptic eyepiece, its big disk seems to float in front of dim field stars. The addition of the OIII brings out definite if not completely concrete hints of the eyes (they always look bigger in the eyepiece than I imagine them to be). They are not overly well defined, but are visible at times, and even occasionally with direct vision.

There's another Messier object in this immediate area, M108, 48′ to the northwest, a dusty almost-edge-on spiral that looks like a smaller version of M82 in small scopes. This magnitude 10.0 galaxy is badly hurt by light-pollution, though, and is most often invisible in all my telescopes.

M3

M97 was OK, but you're tired of the dim? Let's end this tour with a *bang*, M3. This big but bright Globular star cluster, which is found back down in Canes Venatici in the far southern part of that constellation near Coma Berenices, is one of the true treasures of the northern sky. Bright at magnitude 6.3, "medium" in concentration at Shapley-Sawyer Class 6, and large at 18.0′ in diameter, this thing looks good in *any* scope. If you can use a 12-inch or larger instrument on it, you will almost forget you're observing from the city.

You'd think that M3, bright and prominent as it is, would be easy to locate. Surprisingly, this is not always the case. It's in the empty void between Bootes, Canes Venatici, and Coma Berenices, and it is not immediately obvious where to start when

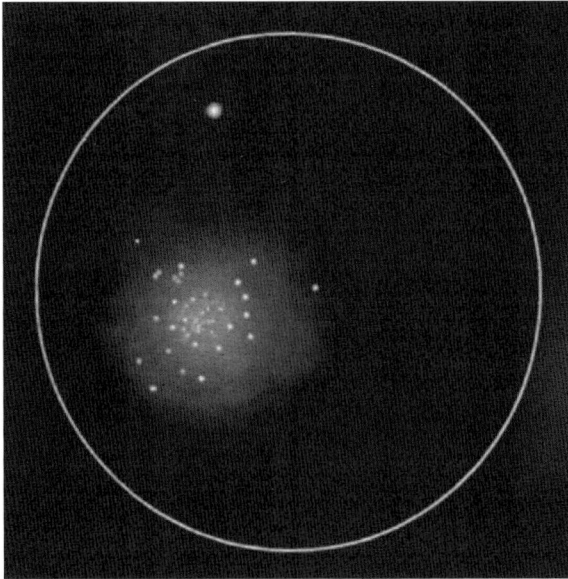

Figure 6.3. The great globular M3.

star-hopping to it. What I do is use brilliant Arcturus (Alpha Bootis) and Rho Bootis, which, at magnitude 3.58 is not as obvious in my skies as blazing Arcturus, but is easy enough to see at any time. M3 forms a near equilateral triangle with these two. If you can't seem to hit it by moving from Arcturus and Rho, another way to approach M3 is by drawing an imaginary line from Rho Bootis to Beta Comae. M3 is two-third of the way along this line in the direction of Beta and just a little to the south of the line. Do whatever you have to do to find M3. It is *breathtaking*. It is an enormous ball of tiny, tiny stars; my drawing in Figure 6.3 doesn't even begin to hint at its majesty. Stars are everywhere, spilling out of the field of a medium-power eyepiece:

> MAN is M3 beautiful! 127× with the C11 reveals many tiny stars from the outer periphery of the cluster and extending inward right across the core. As is often the case with globular clusters, M3 displays hints of color tonight, especially when I lower the magnification. In this case, the general hue is a very subtle blue–green (nearby M13 looks yellow to me). I'm embarrassed to spend so much time on this "easy" object, but it just looks so good!

Don't have a 12-inch scope? Try high magnifications with smaller instruments. That will allow at least some of the cluster's outer suns to wink into view, even in surprisingly small apertures. I've occasionally picked out a cluster star or two with my 80-mm Short Tube refractor, a pretty satisfying feat with such a small scope under such bright skies. Never stop trying until you've squeezed every last photon your telescope, large or small, can deliver from every last DSO.

Tonight's Double Star: *Cor Caroli*, Alpha Canum Venaticorum

Cor Caroli, "The Heart of Charles," as mentioned earlier, is a beautiful double star for even the smallest telescopes. This easy to separate pair is composed of a magnitude 2.9 primary star whose companion, a magnitude 5 bluish-white sun, lies a generous 20 arc seconds away. While the secondary is actually blue in color, the presence of the much brighter primary leads to interesting contrast effects that tend to make the dimmer star take on false shades of purple or "lilac." The star's name, Cor Caroli, was bestowed on it by Sir Edmond Halley (of comet fame) in honor of his monarch, England's Charles II.

> Still haven't had enough? There are many, many wondrous galaxies in this neighborhood, especially in nearby Coma Berenices. Have you seen The Blackeye Galaxy, M64, lately? Doesn't it deserve another look before the season ends? Or galaxy M98 (a hard one)? Or how about M99 the "last" Coma galaxy before you step over the border and into mind-bending Virgo? But if you've had a surfeit of wonder for one night and are ready to quit, don't feel guilty. You can take heart in the fact that these lovely galaxies and their kin will be back tomorrow night and next year. We'll also be back, standing, like Newton, on the shores of a great, dark cosmic ocean, picking-up a bright pebble here and there that our fellow men ignore.

Tour 2

Lion's Den

The spring constellations don't inspire awe based on their appearances. Virgo, Canes Venatici, Bootes, Coma Berenices, and the rest all hold beautiful treasures, but the stars that outline their "stick figures," the traditional patterns of spring constellations, are relatively dim and unmemorable. With one glorious exception. As spring begins in the Northern Hemisphere, Leo the Lion sprawls across the eastern horizon at sundown, a monarch surveying his realm. He's unmistakable, with the backwards question mark or sickle that forms his "mane" being identifiable to even casual sky watchers. He's almost the Orion of springtime.

Leo's not just interesting as a constellation figure; his rising heralds the return of the galaxies. In the spring, the change in seasons shifts our evening point of view from the Milky Way, which is well placed in the night skies of summer and winter, to the wild realms outside our friendly home spiral galaxy. In winter and summer we're looking inward toward the plane of the Milky Way, and its inhabitants, star clusters and nebulae, dominate our evenings. In spring, however, we're looking upward and outward into the darkness of intergalactic space beyond (the North Galactic Pole is found in Coma Berenices). So, what you can expect in spring is galaxies, lots of them. The immense body of our own galaxy is out of the way and its huge clouds of dust and hordes of stars no longer obscure the outside universe. Oh, you can see a few galaxies from the city in summer and winter, but nothing like the riches presented for small scopes in springtime.

When it comes to galaxies, nothing beats the Coma–Virgo cluster. The heart of this great agglomeration of island universes lies only about 60 million light years away (our own spiral is actually an outlying member of the Virgo cluster), and many showpiece galaxies can be found in the Coma Berenices–Virgo area of the sky. Amateur astronomers who enjoy galaxy hunting and observing eagerly await the return of this great cloud every year. I enjoy observing DSOs of all types, and don't really have a favorite variety, but when a new year begins I always find myself longing for the *real* depths of space. I'm ready to step off into the intergalactic void represented by Virgo and the numberless galaxies sprinkled across her like grains of sand. But it seems to take forever for the coy Maiden to reappear in our skies after a hard winter.

Luckily, Virgo is not the only spectacular expanse of galaxies on display in the spring. Long before the Virgin comes back, I'm observing galaxy after galaxy in Leo, who precedes Virgo and is well up over the horizon on late February evenings. The lion is nowhere near as rich in DSOs as Virgo and Coma, but he has enough galaxies to keep me working night after awe-filled night, even from the worst of the light pollution and with the smallest telescopes.

Which is not to say that springtime galaxy observing is always easy or rewarding. Unfortunately, the coming of Leo coincides with the return of stormy weather in many parts of the Northern Hemisphere. As was mentioned earlier, what we need for easy galaxy observing is, in addition to dark skies, *dry* skies. It seems ironic that the best season for touring the dim and distant destinations outside our host galaxy should also be the season when they are hardest for many of us to see.

I've often headed out into a pleasantly warm night, visions of spirals in my head, only to be at least partially stymied by the humid hazes and "isolated thundershowers" that so often mar spring evenings in the Southeastern United States. Sometimes it gets so bad that I find myself thinking that trying to do deep sky observing from my light-polluted backyard is foolish at best. Then, when I'm about to give up and pack my poor, photon-starved telescope away, I begin to have successes that seem almost miraculous considering my usually horrible spring conditions.

One recent April evening I found myself hungry for the deep sky, and especially for galaxies. A quick look outside revealed that conditions were only fair at best. "What the heck," I thought, "if I can't see any galaxies I'll look at Jupiter." Outside with my little 4.25-inch Newtonian reflector—I didn't think there was much reason to bother with one of my larger scopes—I immediately turned to Leo, who, in his accustomed way, was dominating the eastern horizon. Some of his prominent stars were hard to make out in the haze, though, and I supposed I might not see a single spiral.

I pressed on anyway, and after a few minutes hunting, there was M66 peeping in and out. With a little slewing around and averted vision, I found I could see the neighboring spiral, M65, too. M105 was even easier. The sky was far from dark, but still the galaxies were there, perhaps helped slightly by my steady spring seeing. Most veteran deep sky observers know dark, dry skies are critical for optimum observing, but fewer are aware that seeing—atmospheric steadiness—is almost as important for dim-fuzzy-hunting as it is for planet watching. Under steady seeing conditions, extended objects like galaxies and nebulae are stable and not "smeared out" and stars are "smaller" and look brighter.

One thing is sure, after that hazy night I never assumed anything about sky conditions again. Unless you're completely clouded out, always give the sky a try; you might well be as surprised as I was on this "bad" night when galaxy after galaxy landed in my field. After this evening of productive galaxy viewing with a minimalist telescope, the galaxies of Leo never seemed dim and daunting again. Instead, Leo became a very friendly and familiar old lion for me.

Sad to say, this nice lion suffers from several annoying afflictions: lice in his mane, fleas on his stomach, and ticks on his hindquarters. What makes him a very *interesting* lion, though, is the fact that all these bugs are in reality *galaxies*. Referring to these majestic star systems as vermin may seem a little cavalier, but Leo's spirals do conveniently divide themselves into three rather distinct groups, one in his mane (the sickle or "backwards question-mark"), one in his stomach (the area between the stars 52 and 53 Leonis), and one in his hindquarters (the triangle of formed by the prominent stars Theta, Delta, and Beta Leonis).

M65 and M66

M65 and M66 are Leo's best in my opinion, cutting through urban haze with amazing ease. They are not hard to find, either. These two giant spirals are located in the hindquarters asterism about 2° 30′ southeast of Theta Leonis (Figure 6.4). Another guide is magnitude 5.3 73 Leonis, which should be visible in a 50-mm finder, even

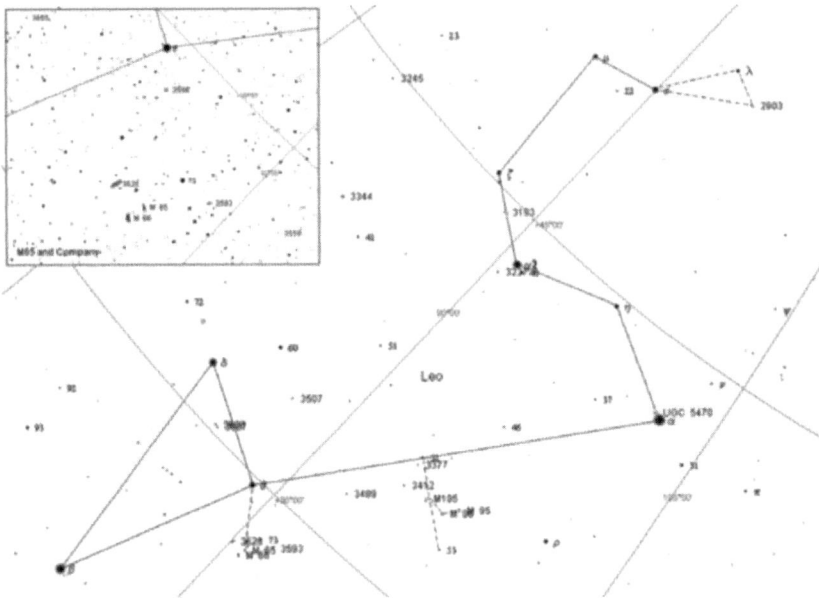

Figure 6.4. Lion's Den.

on badly compromised evenings. M66 is a mere 46′ east of this star. Put 73 on the western edge of the field of a low-power eyepiece and you may see M66 gleaming on your eyepiece's eastern field edge if your telescope can deliver 46′ of sky in one bite. Depending on your sky, you may have to use relatively high magnification to see it easily, but, even then, only a little hunting around in the area of 73 Leonis will turn up the galaxy. Once you have M66 centered, M65 is easy, since it's only 20′ further east. In lower power eyepieces, M65 and M66 should be easily visible in the same field in most scopes if your sky background is dark enough (use a wide-field eyepiece of medium focal length, if possible).

I'm always amazed at just how easy it is to find this pair in the city. If I can't hit them immediately by working from memory, a quick look at a chart like the one in Figure 6.4 makes quick work of the pair. I'm also continuously surprised at how obvious they are. I've often scoffed when I've read descriptions of galaxies that refer to them as "bright"—any galaxy is *inherently* a dim object—but these two are so prominent in my often terrible spring backyard conditions that I guess they really *should* be called bright. To me they both appear more prominent than even their fairly impressive (integrated) visual magnitude values of 10.2 (M65) and 9.6 (M66) suggest.

Once you've found these two, what exactly is there to see? Depending on your eyes and your experience, you may find your first glimpse of them in the city a little disappointing. Two smudges of light, each 3—4′ across. Nothing more. As always, a good, long look can make a difference in the details you can pick out, and it doesn't necessarily take a large scope. In my 4.25-inch Newtonian after extended observation I recorded that they were:

> Lovely and awe-inspiring. A wonderful sight even in this aperture. M66 is the brighter of the two with some hint of a core visible. M65 is dim, but easily seen.

My log entry for this evening also goes on to say that both galaxies are obviously elongated, rather than being just two round blobs, which, unfortunately, is all you see of many galaxies from the city.

Sometimes, throwing more aperture at a DSO helps a lot in the city. Sometimes it doesn't. On a comparable evening—clear, but with haze and high humidity—M65 and M66 looked about the same in my Nexstar 11 SCT as they did in the 4.25-inch Newtonian. They were easier to see and brighter, but gave up no more detail than in the little telescope:

> These galaxies, and especially M66, are fairly impressive in the C11. No core noted for M65, it's an oval smudge of light. M66 is brighter, but looks much the same. The real attraction under these conditions is that both can be seen in the same field of a 22-mm Panoptic eyepiece at 127×.

Actually, even under dark skies, most observers, even those with large instruments, don't see much more than a subtle brightening toward the middles of these spirals. They do look larger from the country, with more of their nebulous outer envelopes visible, but even under the best skies they seem reluctant to give up their subtle spiral arms to visual observers.

M65 and M66 (Plate 29) are large and normal-appearing spiral galaxies, both lying about 35 million light years away. These two are actually neighbors in space, and not just superimposed along our line of sight, but gravitational interaction does not appear to have caused any disruption to their elegant spirals. M65 is classified Sa, while M66 shows signs of a vaguely defined central bulge and therefore is an Sb.

NGC 3628

Let's say you're having one of those often-rare good nights in the spring, dry and relatively dark. You can then try for a "triple play." NGC 3628 lies 30′ north of M66. Although its magnitude value of 10.3 is not much dimmer than that of the other two galaxies, it's considerably larger, about 12′ in extent in photographs. This makes it a good deal more difficult to find, and I haven't seen it from the city in a scope smaller than 8 inches—and that on an exceptional night. Under city lights, a 10–12-inch scope makes seeing this one easier if not *always* trivial. Most of the time, even with the larger scopes, I just barely make it out, as in this log entry from one sultry and smoky June evening:

> This third member of the "Leo Trio" is substantially harder to see than either M65 or M66 in the C11. It's a dim smudge that fades in and out as the seeing changes. Some hint of its strong elongation, but that is just on the edge of perception tonight. Mostly, it looks like a small, dim, round haze. Best seen at 100×.

In long exposure images, NGC 3628 is bizarrely spectacular, showing signs of gravitational disruption. It's seen edge-on and shows a warped equatorial dust lane. It is assumed by astronomers that this disturbed appearance must have been caused by NGC 3628's interaction with nearby M65 or M66 or both. It's puzzling, though, why the other two galaxies look so placid and normal. Like M65 and M66, NGC 3628 doesn't reveal anything much to visual observers other than its basic form. A large scope under country skies shows it as a pretty and bright band of light, but the lovely dust lane and nuclear bulge remain invisible.

The Rest of Leo

Mane Lice

NGC 2903

This Hubble Type Sb spiral shines at magnitude 9.1, but is large at 12.6′ along its major axis, so I expected it would be difficult from the backyard. Not at all. Not in the C11 or my 12.5″ Newtonian, anyway. That this object was always visible in these scopes, even under poor conditions, indicates that it should often be visible in much smaller instruments. I have seen it (barely) in an 8-inch scope, but have not been able to hit it on a night good enough to show it up with less aperture. NGC 2903 can be located by moving 3° 53′ southwest of bright, magnitude 2.98 Epsilon Leonis, the "last" or easternmost star in the mane/sickle. The galaxy forms a tall and slightly skewed triangle with Epsilon and magnitude 4.3 Lambda Leonis, which is 1° 28′ northwest of the galaxy.

My log for a typical spring evening (I recorded fair seeing and some haze) reports that NGC 2903 was

> Visible but not starkly apparent in the C11. Its large disk tends to wink in and out of view as I switch between averted and direct vision. Averted vision seems to show a tiny nucleus at 127×, but I'm not sure of this. Higher powers darken the field but don't help much with the galaxy.

NGC 3190 and NGC 3193

This pair is even easier to find that NGC 2903, since they are located almost directly in the center of a line drawn between two bright sickle stars, Gamma and Zeta Leonis. Although these galaxies are relatively small at 5.5′ and 3.5′ in size, respectively, they are forbiddingly dim at magnitudes 11.9 and 12.4. I have only been able to pick them out with any certitude with 12-inch and larger instruments. You, of course, may have better luck, better eyes, or better skies than I have, so be sure to look for them. If you *can* find them, you're in for a treat. Since NGC 3190 and NGC 3193 are a mere 5′47″ apart, they'll appear in the same field in a fairly high-power eyepiece. That's not all: NGC 3190, is one of the *most* attractive of the dimmer galaxies in Leo, as I mentioned in my observation of the two using the 12.5-inch Newtonian:

> This little pair is a real surprise. NGC 3190 is bright, definitely elongated, and shows a small, stellar core. It really "looks like a galaxy," and not just another smudge. NGC 3193, in the same field, is a typical round elliptical, a fuzzy ball. It does, like its neighbor, show off a nucleus, though. A third galaxy, NGC 3185, should also be present in the field, but I've never seen it from light-polluted home.

Tummy Fleas

M105

M105 has long been one of my favorites, though I'm not entirely sure why. It's a Class E1 elliptical galaxy, so there's no hope of seeing detail of any kind beyond a bright nucleus—all elliptical galaxies are featureless balls of stars without dust lanes or spiral arms to make them really interesting. Maybe it's because this one is very bright and not alone. M105 forms a beautiful little group with NGC 3384 and NGC 3389, both of which are only about 10′ away.

Unlike most of the other galaxies in Leo, M105 and company can be a bit tricky to find, as the only bright stars in the area are 52 and 53 Leonis at magnitudes 5.48 and 5.34, 1° 38′ and 2° 4′ away, respectively. Both stars will likely often be invisible to the naked eye in bright skies. They will appear easily enough in a finderscope, though. Once you've got them located, simply move your telescope to a position midway along a line between the two. You may have to sweep around with the telescope to pick up M105, but it's hard to miss. On any acceptable evening, the galaxy should appear modestly bright in a medium-power eyepiece.

M105, at an integrated magnitude of 9.6 and a size of 4.8′ across, is simply not a challenge at all once you're able to locate it. I've seen it without effort in my 60-mm ETX on clear evenings. The other two are harder, much harder, with magnitude 10.0 NGC 3384, an E7 elliptical, being available to 8-inch scopes regularly and 6-inch scopes on above average nights (with high magnification). The third member of this grouping, NGC 3389, a magnitude 12.0 Sc face-on spiral, is, like other examples of this class and orientation, a "toughie," and usually takes 10–12-inch telescopes for positive identification from urban sites. On poorer nights, it is completely undetectable, even in the 12.5-inch scope. Since the M105 group is a favorite spot in Leo for me, I've drawn it numerous times. My sketch in Figure 6.5 reflects its appearance on a better than average evening. M105 is an old friend now, and I've been coming back to it ever since I rediscovered it with the 12.5″ Dobsonian one fine April evening:

> This trio was quite a treat. I remembered the appearance of M105 fairly well, though I probably hadn't looked at it in years, but I didn't recall the other two galaxies in the field. M105 is bright and round with a stellar nucleus. NGC 3384 looks larger and dimmer than M105 and shows some elongation. NGC 3389 is smaller and a little difficult in the 12.5-inch scope—it was dim enough that I couldn't be sure exactly what its shape was and whether or not it displayed a core. This is an immensely beautiful field—three nice galaxies for the price of one!

M95 and M96

These galaxies are the runners-up in Leo when it comes to attractiveness. Were it not for M65 and M66, they would be the runaway winners without doubt. They are a little more difficult to locate, but not overly hard to find once you're in their general area. If you can find M105, locating M95 and M96 is a snap. Go for M96 first, as it's closest to M105. Move carefully and slowly 48′ south-southeast of M105 and examine the

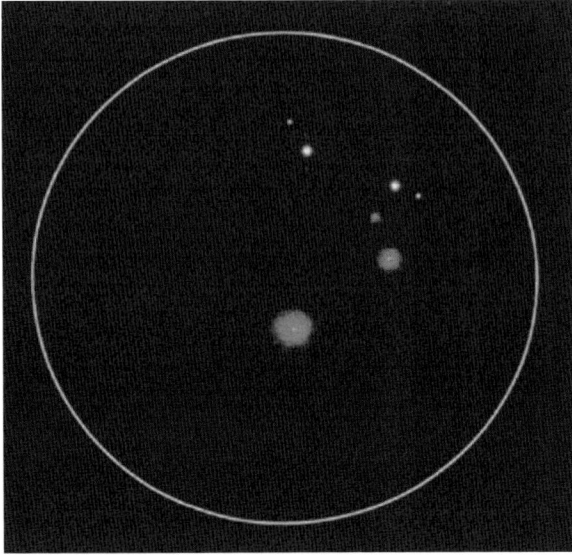

Figure 6.5. M105 and Company.

field obsessively. At an integrated magnitude of 9.3 and a surface brightness of 13.1, M96 will pale when compared to bright M105, but should be visible on most evenings in a 6-inch aperture instrument (as always, depending on the quality of your skies). When you've had a good look at—or for—M96, move on to M95. Its magnitude of 9.7 and size of 6′ × 4′ result in an integrated magnitude of nearly 14, so this object is, not surprisingly, dimmer-appearing than M96. To locate it, eyepiece hop 41′ due east from M96.

Given their close proximity, you'd think these two would make a nice "double galaxy" in a wide-field eyepiece. From dark country skies they certainly do, but in the city you'll need to increase your magnification substantially in order to see either of them. To be successful with this pair, I needed 220× in the C11, and even the field of a 12-mm Nagler wasn't quite large enough to include both galaxies at this magnification. Don't let the above put you off. M95 and M96 are extremely nice despite their relative dimness, being more interesting than many other galaxies in Leo, and shouldn't pose many problems for a 6-inch scope if you work slowly and methodically. They were, in fact, interesting enough to move me to document them in a drawing (Figure 6.6). My log entry notes them as a doable if not easy catch for the 8-inch scope (I was barely able to see them in the 4-inch scope on most nights):

> Conditions are not good and are getting worse as the night wears on, but I didn't have much trouble with this pair. In the 8-inch f/5, M96 is large and fairly prominent. It is obviously elongated and shows a stellar core. M95 is considerably harder and requires averted vision at times, but I can see that it is also elongated and also that it doesn't possess an obvious nuclear region.

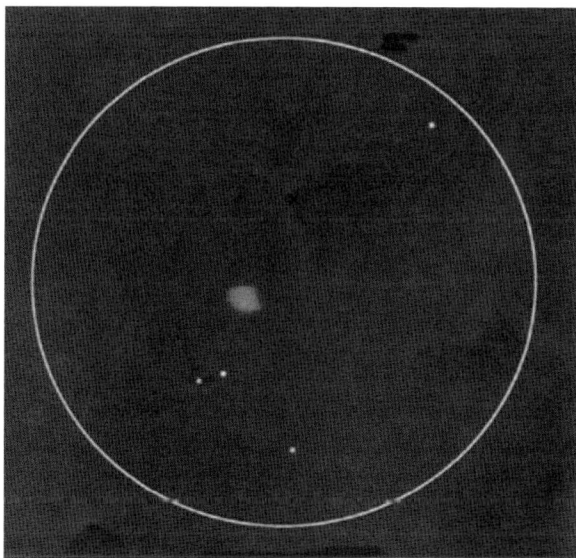

Figure 6.6. M95.

Hind Quarter Ticks

NGC 3521

M65, M66, and NGC 3628 obviously steal the show in the eastern part of the lion, but they aren't the only deep sky wonders to be seen in this area. NGC 3521 is exceptionally beautiful, and I'm surprised that more observers don't talk about it. At magnitude 9.7 and 9.5′ across its major axis, it's easy enough to see in 8-inch and larger telescopes under above average city conditions. In the eyepiece, this SB spiral looks so much like a dimmer twin of M63, the Sunflower, that I've christened it "*Sunflower Junior.*"

Sunflower Junior lives in the eastern section of the Lion, but isn't co-located with M65 and M66 in the triangular hindquarter region. It's in a nondescript area to the south—a full 15° 36′ from Theta Leonis. Forget using any of Leo's bright stars to track down Junior. If you don't have go-to or setting circles to aid you, your best bet is probably magnitude 4.47 Phi Leonis. You should be able to locate Phi with your finder with fair ease, as there aren't many other even marginally prominent stars in the region. To find the galaxy, you'll need to move 4° 31′ to the west of the star. If this doesn't work for you, try starting from the slightly dimmer star, magnitude 4.74 61 Leonis. The trip from 61 is shorter, with the galaxy lying 2° 38′ to the north-northeast.

In long exposure photographs, NGC 3521 really does show a clumpy, many-armed appearance reminiscent of that of the Messier object. Visually, it's remarkable, if a little frustrating:

On this not-so-good night, I was surprised to find NGC 3521 without much of a struggle. At 220× in the C11, it is large, obviously elongated with a stellar core, and its disk seems to occasionally give up fleeting hints of detail, as if a multitude of spiral arms is just on the edge of detection. This detail is on the bare threshold of visibility, however, and I can't hold it or even formulate a good idea of what I'm seeing.

Tonight's Double Star: Algieba

Algieba, Gamma Leonis, is one of the Lion's most prominent stars. It is also one of the finest doubles in the northern sky. This binary pair is composed of a magnitude 2.2 primary star and its magnitude 3.5 companion. They are fairly close together at 4.4″ apart, but, being roughly comparable in brightness, they are not hard for small telescopes to split. At higher powers, the two resemble yellow cat's eyes glowing in the darkness. You can locate Algieba by moving up Leo's sickle from Alpha Leonis, brilliant Regulus. Your target is the second sickle star after Regulus.

What have I taken away from my many nights of galaxy hunting in Leo? Other than the pure exhilaration of viewing these far-away night-dwellers with my own eyes through my own beloved telescopes, they have helped teach me that deep sky observing under city lights is a skill. By trying hard on difficult objects, like the dimmer galaxies in the Leo group, I've trained myself to be a better observer. The next time I have a go at Leo, it's always easier to find my way and find my galaxies. Under dark skies I conquer galaxy after galaxy with laughable ease with a non-go-to non-DSC equipped telescope. I've also learned to be very hesitant about calling an object "impossible" from the city. I succeeded more often than I failed, even on the worst nights, and was rewarded with unforgettable encounters with these distant giants.

Tour 3
The Tresses of Berenice

If Leo is the toe-in-the-water constellation when it comes to galaxies, Coma Berenices is where you step off into the deep end of the pool. This is where the great Coma–Virgo Cluster of Galaxies begins. The southern part of Coma, especially, is almost as island-universe-rich as the depths of Virgo. It's just as interesting, too—or maybe even moreso. Virgo's galaxy fields are impressive, but seem to me to be dominated by less interesting Hubble Types. In my mind, the prototypical Virgo galaxy is a monstrous, round elliptical. Bright? You'd better believe it, but without a trace of the details we strain to see in our city-limited scopes. Coma is a different story. Beautiful (if challenging) face-on Sc spirals mingle with razor-thin edge-ons. Everywhere I turn it seems I find another showpiece.

If you're a beginner, you've probably at least heard fellow amateurs talk about the wonders of Coma Berenices, and have seen the beautiful images of the constellation's many galaxies that appear in the spring issues of the astronomy magazines, but you may not actually have seen the constellation itself. While it's very beautiful, it's quite subdued in an urban setting, since the brightest star in the constellation, Beta Comae, shines weakly at magnitude 4.26 (as you may know, due to Bayer's "mistakes" and changes in the layout of the constellations over the years, "Alpha" is not always the brightest star in a constellation). And what's a "Coma Berenices," anyway? It's the *hair of Berenice.*

Berenice, in legend if not necessarily fact, was the beautiful golden-haired wife of King Ptolemy of Egypt. Apparently their marriage was a happy one, and Berenice pined for him when he was called away to war. She prayed mightily to Aphrodite (the Ptolemys were the Greek rulers of Egypt) for his safe return, swearing she would cut her long and lustrous hair as a sacrifice if he were returned safely to her. When Ptolemy returned home, Berenice dutifully had herself shorn. She must have been inconsolable over the loss of her beautiful locks despite being reunited with her hubby. A court astrologer took pity, however, and pointing at a lovely splash of stars in the spring sky, he informed her that in honor of her sacrifice and devotion, the gods had immortalized her tresses in the night sky.

A pretty story, but most people with some familiarity with the sky, if not Coma, would guess that the constellation in question looks nothing like the glimmering hair of pretty Berenice. That's usually the case when it comes to constellations and mythology—Bootes looks like a kite, not a herdsman, Virgo doesn't look much like a Virgin, and Canes Venatici is two dully shining stars rather than a pair of sprightly hunting dogs. This is one time, though, when a constellation actually resembles its namesake. True, Coma Berenices' basic pattern, located just south of Canes Venatici, is unremarkable. It's just an "L" shape of stars consisting of Alpha, Beta, and Gamma Comae. Unremarkable save for one thing: Berenices hair really *is* entangled in the constellation's pedestrian stars.

What saves Coma Berenices from obscurity as a constellation is Melotte 111. This is a huge cluster of stars 5° in diameter located in the Western half of the constellation, just South of Gamma. From rural skies, the beauty of Mel 111 is incomparable. With an integrated magnitude of 1.8, and its brightest stars hovering around magnitude 5, it's even detectable in the city on a nice night when Coma is riding high in the sky. Because of Mel 111, Coma Berenices puts nearby Canes Venatici to shame, and is more readily identifiable even than sprawling Virgo, whose only claim to fame is her bright sapphire, Spica. While Mel 111 is often completely invisible to the naked eye under city lights, a pair of binoculars easily brings its myriad stars out of the gray sky background. Before going on to this area's hard-core objects, take a few minutes to admire this cluster with a pair of 10 × 50 binoculars.

Learning constellation lore is fun and Mel 111 is lovely, but I know you're here for the *meat*, the deep sky objects (DSOs) mingled with our lady's shining hair. There's no lack of them, and I could go far beyond what's outlined here, but these are the best of the best. Where to start? Make it easy on yourself in two ways: start with a Messier and begin with the less galaxy-crowded northern half of the constellation.

M64

M64, an integrated magnitude 8.5 Sb spiral galaxy, is the perfect place to begin our leisurely stroll through Coma's galaxy fields. Under dark skies you'll see galaxies almost anywhere in Coma Berenices where you point a medium-aperture scope, just as in Virgo, and playing "which galaxy is which" gets confusing in a hurry. The area of M64, located to the north of the Coma–Virgo border, back in the direction of the constellation's "L" shape, is, thankfully, a less crowded and confusing part of the sky. In fact, M64 is alone in its surrounding patch of the heavens for the urban astronomer.

Finding M64 is incredibly easy. It is about two-third of the way along a line drawn between magnitude 5.6 40 Comae and magnitude 4.9 35 Comae. With your atlas or Figure 6.7 properly oriented, both of these stars should be easy to locate with a large aperture finder—like Virgo, this part of Coma is relatively barren in light-polluted skies, so there aren't hordes of stars to confuse a galaxy hunter. Bright enough to appear fairly obvious at 50×, M64 will initially appear as a dimly glowing oval of light a minute or two of arc across.

Were it merely an oval of light in our telescopes, M64 would be just another spring galaxy to be glanced at quickly before moving on—for those of you who can bring yourselves to be blase' about these great cosmic creatures. But there's one curious feature of M64 that puts it in the "must see" class. M64 has a nickname, you see, The *Blackeye*. This is due to a large patch or lane of dust located just outside the galaxy's nuclear regions (See Plate 30) In photos, this looks more like a curving arc than the spot or "eye" we see visually in small to medium size telescopes. How easy is the galaxy and its eye? M64, whose true dimensions in photographs are 10.3′ ×5.0′ across, is not a challenge for most urban observers. With sufficient magnification even a 60-mm can hope to pick it up in the worst light pollution a mid-sized city can throw at it. How about the dust patch, the eye? That is a different story, as I found out with my 6-inch $f/8$:

M64 is easy as pie in the 6-inch at 96× with a cheap 12-mm Kellner eyepiece. Certainly not overwhelming, though. Compact (but not stellar) core surrounded by sharply defined oval haze oriented roughly northeast-southwest. I convinced myself I saw evidence of

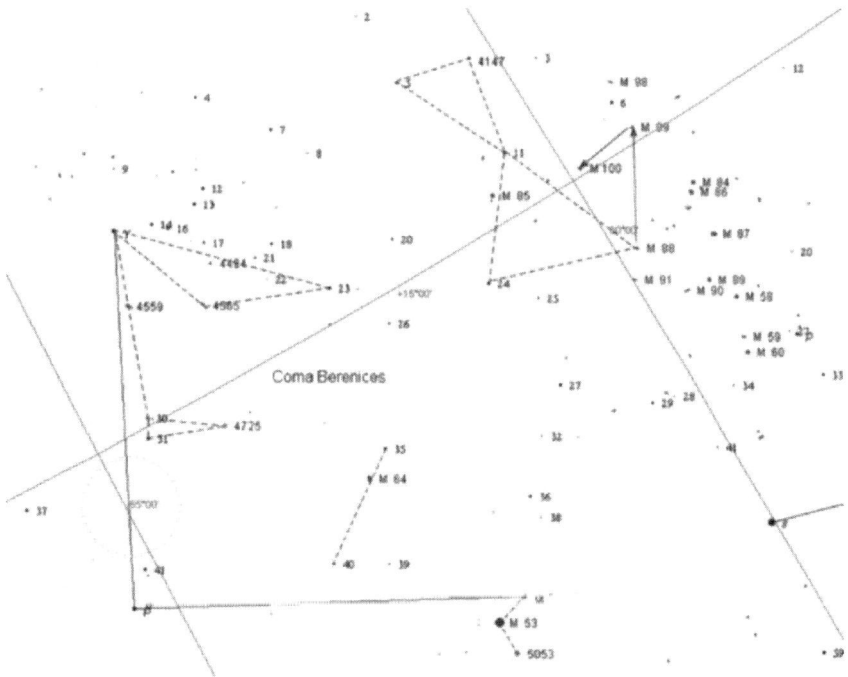

Figure 6.7. Coma Berenices.

the black eye, but, in truth, I'm not sure if I saw it or not. It's incredibly subtle in this aperture in the light pollution, and I think steady seeing in addition to dry, transparent skies would be a big help.

As is usually (though not always) the case with galaxies in the city, going to the C11 SCT helped. A lot:

> The famous "black eye" of M64 is immediately and dramatically obvious. On this above-average night, the galaxy itself is at least 3′ across, and the black eye appears as a notably darker area next to a bright patch directly adjacent to the stellar-appearing nucleus. Though the eye is easy enough to see, I found that I had to use averted vision to get a good look at it most of the time, and even then could not always "hold" it. This is not an easy feature for a city scope, even an 11-inch scope. I used 220×, but something closer to 300× might yield better results.

M64, the Blackeye (I've also heard this object referred to as The Sleeping Beauty Galaxy—don't ask me why), is a large and healthy spiral, a member of the Coma Supercluster of galaxies, and lies around 19 million light years from Earth. "Healthy" as in "not disrupted." But M64 is not *quite* your normal galaxy. It appears to be composed of an inner and an outer disk of stars that are rotating *counter to each other*. The interaction between these two disks is probably responsible for the vigorous star formation we see taking place and also, perhaps, for the weird eyespot. It's speculated that the whole shebang may be the result of M64 having devoured a smaller, but still sizable, companion galaxy many millions of years ago.

NGC 4565

M64 is a "showpiece," but it's not *the* showpiece of Coma. That honor must go to NGC 4565. In medium-aperture scopes, it is one of the most beautiful objects in the entire sky. This near-edge-on magnitude 9.6 Sb spiral is composed of a bright, round central region 2 or 3′ in extent superimposed on the thin disk of the galaxy that extends for an amazing 15′ of arc (Plate 31) This domed-disk look gives NGC 4565 its nickname, "The Flying Saucer Galaxy."

Getting there is easy, if not quite as easy a trip as M64. Your guides to this one are magnitude 4.4 Gamma Comae, and magnitude 4.8 23 Comae, which lies 6° from the brighter star. NGC 4565 is almost exactly halfway along this line, and 1° 45′ northeast of it. In other words, Gamma and 23 form the base and NGC 4565 the apex of a shallow triangle. Another way to pin down NGC 4565 is by looking for magnitude 5.3 17 Comae. NGC 4565 forms a right triangle with this star and Gamma, and is 1° 39′ north-northeast of 17. Actually, when it comes to locating the galaxy, you have an advantage over those astronomers blessed with darker skies. For them, this area is a riot. It's just *brimming* with the stars of Melotte 111. For you, under the sodium streetlights, 23 will stand out pretty prominently.

If you've seen this one from the country, you'll remember it as astonishingly big and surprisingly bright, even in 4-inch instruments. But the question is always, "How much galaxy can I see in the city?" Aperture is a tremendous help. In urban 4–6-inch instruments, hunting NGC 4565 can spell frustration, with the galaxy sometimes being completely invisible. In an 8-inch scope, it *can* be fairly easy to see, but much of its dark-sky glory has been stripped away. Instead of being a bright edge-on streak, NGC 4565 may be reduced to just another Virgo–Coma Fuzzy Ball. If you're looking for the famous Flying Saucer, you may even pass over it, as you're expecting a distinctive spindle-shape, rather than just a round nuclear region. On very good nights with an 8-inch scope, or on average nights with a 10-inch or larger scope, the saucer shape can be easy, however (Figure 6.8):

> Plenty of skyglow tonight, though I would rate this late spring evening as drier than average. With direct vision at 127×, NGC 4565 first appears as a nebulous round blob about 1′ or less in diameter with a tiny, bright, star-like nucleus. A little averted vision quickly reveals the edge-on disk that forms the Saucer. With continued examination, I'm confident that I'm seeing at least 5′ of disk on either side of the core—pretty impressive for a C11 in light-pollution hell!

Note that while I was able to detect the saucer's disk with the 8-inch scope on good nights, I was never able to convince myself that I saw real evidence of NGC 4565's equatorial dust lane, the equivalent of the "Great Rift" that cuts through our own Milky Way's plane, and which is so obvious in photographs of the Flying Saucer. I tried for it with all of my might and all my telescopes from the wide-field 60-mm up to the 12.5″ Dobsonian at a variety of magnifications, but failed miserably. Which doesn't mean you might not succeed, or that I might not bag it on a special night.

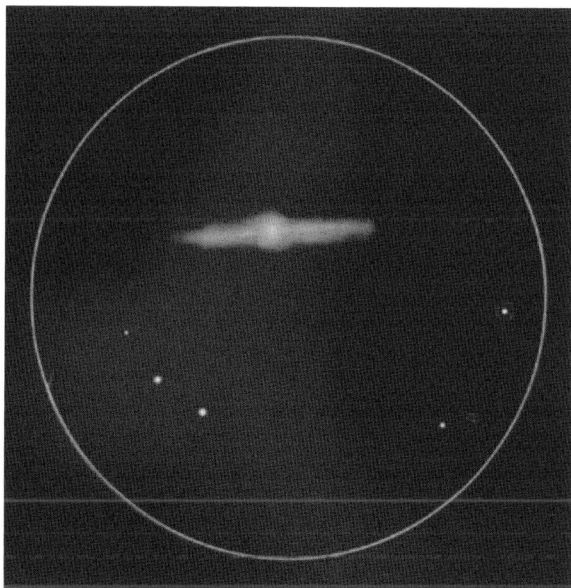

Figure 6.8. The magnificent Flying Saucer Galaxy.

What a sight NGC 4565 must be from close-up. It's the largest edge-on spiral as seen in our telescopes, and, depending on whose distance estimates you use, it is also large in reality, ranging up to a diameter of 125,000 light years, considerably larger than the Milky Way, which is probably no bigger than 90–100,000 light years in diameter. This object, missed by Messier, was first plucked out of the void by William Herschel in 1785, and has been a source of wonder and joy for sky watchers ever since.

M53

Just like dear, old Mother Earth, the Milky Way Galaxy has an axis of rotation and, thus, a North Pole and a South Pole. Looking in the direction of its North Galactic Pole, which is located not far from the unassuming star 31 Comae, you're looking up and out of the galaxy's disk, 90° from the Milky Way's plane. You're peering into deep space, so it's no wonder there are so many galaxies in this area. You'll look in vain for bright nebulae and open clusters as you scan north–south from Coma to Virgo. Globular clusters are a different matter, however. While they congregate along the galactic plane in places like Sagittarius and Ophiuchus, their orbits around the Milky Way's center and our perspective can bring them into view even in the midst of the Realm of the Galaxies.

If there's an easier DSO to find than magnitude 7.6 M53, I've never seen it. Well, maybe the Orion Nebula is a *little* easier to hunt down, but not much. M53 is located

56′ north and slightly east of brilliant Diadem, Alpha Comae. Position the scope in the general vicinity and you'll likely find M53 staring back at you without any hunting at all. If your telescope can deliver a 1° field, put Alpha on the Northern field edge and M53 will "automatically" be visible on the other side of the eyepiece field (once you see it, move Alpha out of the field to reduce the glare, of course). I found the cluster trivially easy to spot even with my 60-mm ETX during typically hazy spring weather here on the Gulf of Mexico coast.

How *good* is it? In the 60-mm, as you'd expect, it was just a little round ball of nebulous light. In the city, a 4-inch scope makes it bigger, but doesn't resolve any stars. Even in the county, this can be a tough nut for a 4 incher to crack. At 6 inches, things get better in a hurry:

> Bright and easy to locate. Round with a grainy, diffuse core at 96× in my 6-inch f/8 Newtonian reflector. As I continue to stare at M53, I'm surprised to see stars start popping out at the "edges" in this medium-difficult object.

Encouraged by this 6-inch scope observation, I threw more aperture and more magnification at M53:

> Very pleasant in the 12.5-inch Dobsonian at 150×. Two fairly bright stars in the field about 15′ from the cluster's center. As the sky darkens, M53 becomes a real mind-blower! Microscopically tiny stars wink in everywhere, including across the core, and M53 assumes a weird "splayed" appearance, looking like a cosmic jellyfish floating among the stars. This impression is heightened by the fact that the core is not overly bright or concentrated.

Swimming alone in the star-poor reaches of Coma's deep ocean, M53 (Plate 32) evokes a sense of lonely isolation, and this feeling does actually reflect the truth of things. M53 is one of the more distant globular clusters known, being, it's estimated, 70,000 light years from Earth. Like other globular star clusters, it is in a huge long-period orbit that will eventually carry it back closer to "home." But for now, M53 sails along distant shores in splendid isolation. This Shapley–Sawyer Class 5 globular's brightest stars are at around magnitude 13.8, which is why it's hard to resolve any stars with small telescopes at anything but high magnifications and under dark skies.

The Coma "What Else"

NGC 5053

What other good stuff is to be seen in Coma Berenices? Plenty. If you like galaxies and have a medium-sized telescope, Coma can provide countless hours of enjoyment, even if you must observe from considerable light pollution. We're into the Realm of the Galaxies here, of course, but there are at least two other objects worthy of our attention. The logical next stop is another globular, NGC 5053.

M53 floats far from home, but not completely alone from our perspective. There is another globular very closer at hand, NGC 5053. It's considerably nearer to Earth than M53, being approximately 53,000 light years from us, but it's along our line of sight to M53 and appears only about 1° from that cluster in the sky. A quick look at its specs: magnitude 9.8, 9′ in diameter, and a Shapley–Sawyer Class of XI (11), and it's obvious that small scope owners are in for a real battle. It's distant, it's relatively dim, it's fairly large, and it's very weakly concentrated, meaning the 4-inch owner probably can't hope to see even an unresolved core. The surface brightness of this object is a distressing magnitude 14.3. Finding is not the problem, seeing is. It's in the same field as M53 with a wide-field eyepiece at low power—but you probably will never see it at very low power in the city. On a good spring night with the C11, I tried my hardest with this one, and was rewarded with at least a look at its dim reaches:

> Some of NGC 5053's small stars occasionally swim into view in the C11 at 127×, and, as the cluster rises toward culmination, I can see quite a few members at one time or another. This cluster is unexpectedly difficult. My impression is that this looks like a very dim, large, and flattened open cluster rather than a globular. No central condensation at all.

I keep trying for NGC 5053 with my smaller telescopes, but so far have been unsuccessful with it from the city. Actually, this globular really didn't look impressive or interesting from the pristine dark skies of the Texas Star Party at Prude Ranch in Fort Davis with the 12.5-inch scope.

M88

Ouch! How about something just a little easier? Now we turn our attention to the heart of the matter, the Coma galaxies near the Virgo border. This is a rich and confusing area, so take a deep breath and get a detailed finder chart ready. Don't be too apprehensive, though. The four Messier galaxies we'll visit are distinctive and bright, so you shouldn't have too much trouble identifying them. Luckily, they reside just outside the really galaxy-congested areas to the South.

M88 provides a good jumping-off point if you decide to eyepiece-hop through the region (the galaxies in this area are close together enough to make this practical). It's an Sc spiral of the reasonable magnitude of 9.6, and appears relatively bright in the eyepiece. Unfortunately, if you don't have a go-to scope, it can be something of a bear to find. The best way to this one is by using two Mel 111 stars, magnitude 4.7 11 Comae and magnitude 5.0 24 Comae as your markers. The galaxy forms a near equilateral triangle with the stars, and is positioned due east of them. When you're in the correct spot, it should show itself in fairly small apertures, though you may need to run up the magnification as high as you dare. Work carefully while hunting this galaxy, referring to your atlas frequently. It's easy to mistake M91 for M88, though M91 is obviously a good bit farther east on the charts. To avoid this case of mistaken identity, print out a detailed eyepiece-tailored finder chart with *Cartes du Ciel* or your astronomy program of choice.

As frustrating as M88 can be to find and positively identify, it has one thing in its favor, it doesn't require extreme aperture for a good look. I know I got a good view of it with my Chinese-made 8-inch *f*/5 Newtonian.

> This is a superior night in the city, and M88 is nice and bright at 167× in the 8-inch f/5. What I see is a prominent core region. This area is large, about a minute of arc across, and is elongated north–south. I also see hints of a small nucleus, but not a trace of spiral arms.

Naturally, I always want to see more detail, and when I came back to this one on a similarly good night with the 12.5″ Dobsonian, I was able to see a faint outlying haze in addition to the bright oval central region the 8-inch scope revealed. This haze appeared as two faint extensions on either side of the nucleus, and I'm confident that this averted-vision view was showing me traces of this spiral's lustrous arms.

M99

Next up is M99, a fairly considerable jump, 3°12′ due west, for the eyepiece-hopper, but easily doable if you keep track of the prominent field stars sprinkled along the way. Unfortunately, M99 isn't easy to see on an average city night. Not in the 8-inch scope. Or really even in the 11 and 12.5-inch telescopes. Why? At magnitude 9.9, it's not much dimmer than easy M88. It's that same old bugaboo we often run up against with galaxies: it's a face-on Sc spiral. As they get dimmer, they get much harder. The only saving grace here is that it's relatively small at 5.3′ × 4.6′ in size, which made it at least detectable in the C11:

> On this good night, M99 is fairly easily visible in the C11. "Easily visible" in that I can make it out as a dim and somewhat formless glow in the field of a 220× eyepiece. Increasing my power to 300× and above gives an impression of elongation, but it's hard to hold the galaxy steady enough in my gaze to be sure.

Naturally, all I was seeing was M99's inner regions. There was not a hint of its delicate spiral arms.

M100

A short hop of 1° 42′ north-northeast brings us to another face-on Sc spiral, M100, nicknamed the "Catharine Wheel Galaxy." If you're not eyepiece hopping from M99, your best landmarks to M100 are 11 and 6 Comae, both of which glimmer at approximately 5th magnitude. The Galaxy is approximately 2° from either star and 1° east of a line drawn between them. M100, at 7.5′ × 6.1′, is larger than M99, and only slightly brighter at Magnitude 9.4, so it should be a little harder to see due to its more spread-out light. In fact, many observers do comment that M100 is more difficult than M99. Not me, though. This galaxy was not only easier than I expected in the 11-inch scope, it seemed to show more detail than M99:

> M100 surprised me—I figured I'd really have to fight the seeing and light pollution to get a look at it, even in a C11. But there it was, showing about 45″–1′ of its inner core. Not only that. Averted vision clearly showed faint haze extending out at least another minute or two of arc. Looked substantially better than M99 this evening.

Under dark skies, there's a very attractive near-edge-on galaxy in the same medium-power field as M100, NGC 4312. Despite this "bonus galaxy" being prominent in an 8-inch scope from dark skies, however, I don't believe I've ever seen it from the city with any of my scopes. From the country, in large-aperture scopes, M100 itself becomes a marvel, with its far-flung arms actually seeming to mimic the sparkling of a fireworks Catharine wheel.

M85

The next Messier on our path is a good meat-and-potatoes galaxy in almost any but the smallest instruments. From M100, move the telescope 2° 26′ north. To find it directly, position your telescope 3/4 of the way along a line drawn between 24 and 11 Comae. The galaxy lies closest to 11, and is 1° 10′ from the star. Once there, the galaxy appears as a little smudge of light at an integrated magnitude of 9.1. In images it's 7.4 arc minutes across its long axis and about 5.9′ in "width," but in the city be content if you can make out an arc minute of galaxy at best. Despite its true elongated shape (this is a lenticular S0 type galaxy), amateur telescopes of almost any size and under any skies just deliver a round fuzzy:

> M85 is prominent enough to be visible with direct vision in the 6-inch f/8 Dobsonian—though I wouldn't exactly call it bright! Moving South in Coma puts me into increasing light pollution at this site. Round with a condensed but not star-like nucleus. A bright field star is present 1′ outside the galaxy's core.

Like M100, the field of M85 harbors a "companion" galaxy, NGC 4394, but, like M100's buddy, this magnitude 10.9 sprite seems to be invisible in the city—in my telescopes, at least. As always, though, I keep trying.

That wasn't so hard, was it? Yes, southern Coma is clogged with DSOs, but as long as you keep your wits about you and your charts at hand, navigation can be easy and even fun. Relax. We're done with this challenging part of the constellation, now.

NGC 4725

For the next two objects we jump back to the northwestern area, the same part of Coma we visited for NGC 4565. At 2° 3′ south of 31 Comae, we find NGC 4725, an oval, magnitude 9.40 phantom glowing bravely despite a size bordering on the large, 10.4′ × 7.2′ of arc in images. This DSO can be slightly tricky to find, since there aren't any bright stars in its immediate area. One tip-off is a prominent (in a 50-mm finder) magnitude 6 star, HR 4864, that is 45′ farther south than the galaxy and almost on a line running from 31 and through NGC 4725.

In the C11, this near face-on SAB galaxy appears as a prominent oval haze 2′ in size. I noted a bright core and strong elongation east–west. Observers with darker skies than mine often report seeing details near the nucleus in NGC 4725, possibly evidence of this barred spiral's arms. I've never seen or suspected that in the city, however.

NGC 4559

This amorphous galaxy is easier than NGC 4725 when it comes to locating, but harder to observe. It's in the Western part of Coma, 2° from Gamma, and about 2/3 of the way along a line drawn between 31 Comae and Gamma Comae, almost directly north of NGC 4565. This is another big one, a strongly elongated SBc barred spiral that stretches 11.0′ × 4.9′ and glows weakly at magnitude 10.5. I didn't expect much, not even with the 12.5″ Dobsonian, so I wasn't disappointed:

> A challenging patch of nebulosity. A small triangle of bright field stars makes this a more difficult object than it would probably otherwise be. Definitely not easy. I think I see a shapeless or maybe oval (?) form a few minutes across.

NGC 4147

Let's finish our evening's walk with an object that's definitely not a showpiece, but which is something of a surprise and considerably more entertaining than NGC 4559. Coma Berenices contains two globular clusters, right? Sure, everybody knows that! Wrong. There are *three*. The third is little NGC 4147, a 4′ smudge glimmering dimly at magnitude 10.3. This faint fuzzy is in the barren southwestern area of the constellation, well to the west of the galaxy packed Virgo border, but it's not *too* difficult to find, since it forms a triangle with 5 and 11 Comae, and is only 2° from either star.

I stuck to larger aperture for this Class VI globular, the 11-inch Schmidt Cassegrain, since I hoped to tease a star or two out of it at high magnifications. It will no doubt be visible in smaller telescopes given its fairly small size despite its seemingly punishing magnitude (for a DSO in the city). I *was* pleased with its appearance, though, try as I might, I could not make any stars appear on this particular evening:

> A little blob of a glob that stands out amazingly well at 127× in the C11. It was very obvious the second I put my eye to the eyepiece after the go-to slew finished. It's a bright core with an envelope of dim nebulosity extending 1′ outward on all sides. Going to magnifications of 300 × −500× did not resolve any stars, but did cause the cluster to take on a grainy appearance, as if it were on the edge of resolution. I'll definitely come back to this one.

I realized from the beginning that it would probably take very dark skies and at least the 12.5-inch scope to pick *any* stars out of NGC 4147, but I'll keep trying, hoping to be proven wrong. That is the essence of City Lights deep sky astronomy.

Tonight's Double Star: 24 Comae

24 Comae is wide and easy with a separation of 20.4 arc seconds between its two stars. What makes this pair notable is that the color contrast between the magnitude 5.2 primary and the magnitude 6.7 secondary results in a case of "false color" for the dimmer star. In reality, the primary is a strong orange and the secondary is white, but, as is often the case with this color combination, your eye and brain insist that the dimmer star must be an amazing emerald green. There are no green stars, but you'll find that hard to believe after observing this interesting double.

The subdued combined magnitude of the pair, 5.2, and its presence among the hordes of cluster stars in Melotte 11 make 24 somewhat difficult to find without go-to or setting circles. Luckily, the star forms a near right triangle with Alpha and Beta, so it's not hard to position the scope on the right spot. The large separation between the primary and secondary stars means you can use a low-power eyepiece, which should turn up this pair of gems with ease.

> Even just hitting the highlights of Coma Berenices leaves me breathless. There's so much here to study and admire. And I'm not alone. Today, so long after the fabled king and queen have returned to dust, sky watchers everywhere still gaze in awestruck wonder at Berenices' shining and immortal tresses.

Tour 4

In the Arms of the Maiden

Virgo is scary. I don't mean the goddess, the Greek goddess Persephone who's some-times identified with Virgo. *She* seems to have been something of a shrinking violet, acquiescing to Hades with only minimal protest, and eating those pomegranate seeds willingly enough. It was her Mother, Demeter, you had to look out for, unless you like cold weather. When I say "scary," I'm talking Virgo the *constellation* and Virgo the *cluster of galaxies*.

When I first began observing the deep sky many years ago, I was eager to navigate the Great Virgo Supercluster. Where else can the urban astronomer confront galaxy after galaxy, many of which are bright Messiers? There are *eleven* Messier galaxies within Virgo's borders. After a few nights with the Maiden, however, I turned-tail and *ran*. I didn't dare the area Hubble called the "Realm of the Nebulae" again for several years. Because I couldn't find any of Virgo's objects in my moderately light-polluted skies? No. Because I found *too many of them*. Even my 4.25-inch reflector delivered faint fuzzy upon faint fuzzy. I was seeing many galaxies, but had no idea *which* galaxies I was seeing, became hopelessly confused and withdrew.

Obviously, I finally made my way back to Virgo and found success there, both in the city and the country. What made the difference? Having the proper tools and a *plan*. Nowhere in the sky is it more critical to have a detailed star atlas. Virgo features a few bright stars like her lustrous magnitude 1.0 sapphire, Spica, but is, overall, a subdued constellation. The area where most of the galaxies reside, within the "arms" of the "Y" in the western part of Virgo, lacks prominent guide stars. During my first attempts at the constellation, I was using the time-honored *Norton's Star Atlas,* which only shows stars down to magnitude 6, and it was almost worthless for the task. In *Norton's*, the area of interest between Virgo's arms has the prominent galaxies marked, yes, but far too few stars are shown to help the star-hopping galaxy hunter pick his steps. Even if there had been more stars indicated, I probably wouldn't have seen them in my paltry 25-mm finderscope. Equipping myself with Becvar's *Skalnate Pleso Atlas* (the forerunner of *Sky Atlas 2000*) and a 50-mm finder made all the difference in the world.

I also had a plan. The secret to navigating Virgo, as laid out in Chapter 3, is to *eyepiece hop* from galaxy to galaxy. Virgo is so crammed with objects that it is eyepiece-hopping heaven. You'll use *Sky Atlas 2000* or another similarly detailed atlas to plan your general itinerary, but your mainstay will be computer-generated charts similar to the one in Figure 6.9. As mentioned earlier, any modern computer-aided-astronomy program can produce roadmaps like this, but if you're a penny-pincher like me, you'll use the free *Cartes du Ciel*. If you don't have access to a computer, a few pages photocopied from *Uranometria 2000* and marked by hand with appropriate eyepiece field circles will work as well. If you are using a PC, zoom-in on the area between the arms, click on the first galaxy of interest, plot an eyepiece circle that corresponds to your setup, and move to the adjacent object on your list. You're making *stepping stones* through the sky—that's what these field circles are. You will also want to adjust the stellar

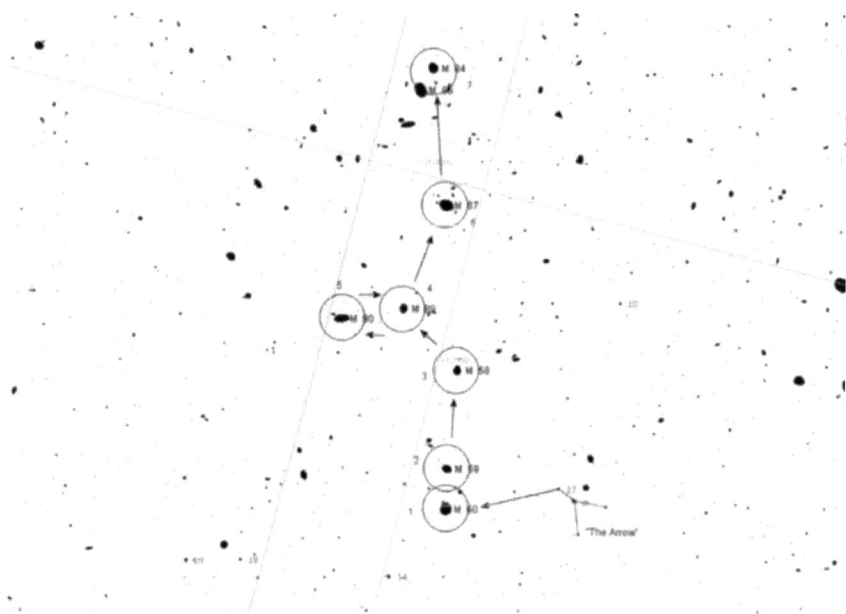

Figure 6.9. Virgo Star Hop.

magnitude settings of the program in order to have only those stars that are likely to be visible with your telescope and eyepiece combination displayed and printed on your charts.

Not sure how deep your telescope and eyepiece will go? Pick an area around an easily identifiable landmark star, maybe Rho Virginis, and print out a series of charts of the star's area overlaid with eyepiece circles that match your ocular and scope setup. Make several charts, each going dimmer in magnitude, point the scope at the area, and see which chart matches up best with what you see in the eyepiece you plan to use while hopping.

Which eyepieces are most appropriate for conquering the Virgin? Review the accessory section earlier in this book, but, if you're on a budget, some of the new imported Chinese wide-field eyepieces are very inexpensive and sport generous apparent fields of view. Newer ones sometimes give you as much as 80° of apparent field, which is perfect for eyepiece hopping. You will also want to stick with an eyepiece that gives you medium magnification—125× was optimum for my skies. If you can't find one of the inexpensive 2-inch wide-field eyepieces with a short enough focal length to give you powers in this range with your telescope, consider a 2-inch Barlow. Imported ones of good quality are now available to go along with the imported 2-inch format eyepieces.

How about scopes? What's needed for success in the Realm of the Nebulae? As we found in our wanderings through Leo and Coma, galaxies are dim. That's just the way it is. On a very superior night in the city, you should be able to do any of the following with a 6-inch scope. On poorer nights, or from really badly light-polluted areas, though, any or all of these *may* be invisible in an 8-inch scope. When I was making my "test runs" for this tour, I chose to throw all the aperture I had at Virgo—12.5 inches. On all but the very worst, haziest nights, the 12.5-inch Dobsonian would reliably bring back all of these galaxies, and would often show me the fainter "companion" galaxies that frequently reside in the same fields as the target objects in this crowded area.

The tour that follows is necessarily only a sample of Virgo, and a small one. It doesn't exhaust the Messiers in Virgo, and many of the multitudinous NGC galaxies marked on your charts will also be visible in light-polluted skies. This selection will show you the technique for eyepiece hopping and will whet your appetite for more of the Great Virgo Cluster.

Before getting started, I want you to know a little more about the area you'll be traversing. It's not mere hyperbole when I refer to the *Realm of the Galaxies* (as we now know Hubble's "Nebulae" to be) as "awesome," "mind bending," or by some equally extreme adjective. This crowd of galaxies deserves all that and more. The Virgo Cluster is a great grouping of island universes that stretches from north to south across our spring skies. By conservative estimates, it contains at least 2000 galaxies ranging from inconsequential dwarfs like our own galaxy's Magellanic Clouds to monstrous heavyweights like the frightening M87, a great elliptical galaxy residing near the cluster's heart and containing, at a conservative estimate, *100 times* the mass of our own (large) Milky Way. The center of the Virgo Cluster is roughly 65 million light years away, and is itself is a component of a larger structure, the Local Supercluster of galaxies, that also encompasses Coma's many members.

The Jumping-Off Point

In addition to good charts, finders, and eyepieces, there's one other critical need for wending your way through all these island universes: a convenient jumping-off point. That is, a star or asterism near your first target that will get you going and to which you can easily return or backtrack to if you get lost. As mentioned in the initial discussion of *eyepiece hopping* in Chapter 3, we'll use a little "Y" of stars in northern Virgo. When I talk about a "Y" in this case, I don't mean the great form of the constellation, but the little grouping of stars centered on Rho Virginis as shown in Figure 6.9. This asterism stands out well in a 50-mm finder and you shouldn't have much trouble getting there from epsilon Virginis. Position that "Y" in your finder, and lets go to town—to downtown Virgo.

M60

Our first stop is M60, a nice, bright magnitude 8.8 Messier elliptical. It's a good place to begin our quest, even if featureless ellipticals like this one are the least interesting galaxies to view under any conditions. If you're properly positioned on the Little Y, finding M60 is a pure joy. Think of the Y as an arrow pointing the way to the galaxies of Virgo. Starting from the "tip" of the arrow, 27 Virginis, move your scope 1° 14′ north and just slightly to the east.

Once you're on the field, you should immediately see at least one dim specter of a galaxy. If you're lucky, you may notice another faint fuzzy. There's a dimmer object, the small, magnitude 11.9 spiral, NGC 4647, there, too, but identifying M60 is not a problem; it is the obviously larger, brighter galaxy. Under fair seeing and transparency on a March night, the 12.5-inch reflector didn't make M60 overly exciting, but it was nice to be successfully started on my way across the Virgin:

> M60 is bright and unmistakable with direct vision in the 12-mm eyepiece. It's about 2′ in diameter tonight, and looks perfectly round with subtle hints of a small, bright core. No hint of NGC 4647 this evening. Two bright field stars make this an attractive area.

Since this is an elliptical, there's little more than this to expect from any telescope under any conditions. With dark skies and large scopes, M60 gets bigger, but is just a bigger featureless ball. Although this is a not-quite-round E2, it's going to be very difficult to make out such a small amount of elongation visually in any instrument.

M60, the easternmost of the Virgo Messiers, may not look impressive, but its stats reveal this as *one big sucker*. At 60 million light years out, its 7.2′ × 5.9′ disk corresponds to a true size of 120,000 light years across. Its mass is estimated at around 60 billion Suns. In photos, the little accompanying galaxy, NGC 4647, appears disturbed, and it is thought that it has had an ancient encounter with M60. In the heart of Virgo, interactions like this are common and often result in the smaller spirals being "devoured" by the great ellipticals, who just get fatter and fatter as the ages pass.

M59

So, you thought M60 was an easy find? The hop to M59 is even more effortless. It is 25′ to the west from M60, so all you have to do is gently nudge your scope, moving slowly west while continuing to look through the main scope's eyepiece. Like M60's field, M59's field also contains two bright stars. These make the field more distinctive, but the galaxy doesn't need any help. It is one of the more interesting and attractive objects on tonight's journey. When you stop your scope, a magnitude 9.6 oval of light should be obvious. While this is an elliptical, it's an *E*5 elliptical, a strongly oval thing well toward the lens shapes of the "late" elliptical galaxies and the S0 lenticulars.

In the eyepiece of a 6-inch *f*/8 reflector, M59 is lovely, if not overwhelming:

> M59 is strongly elongated, a little oval puff of smoke. No core, no outer envelope. Looks like a fingerprint smudge on black velvet. Despite this lack of details, this galaxy is one of the more unusual sights in Virgo. An odd little spot, sharply defined, and standing out from the background incredibly well.

M59 is another "giant" elliptical, if not quite as large as neighboring M60. It's 90,000 light years in extent along its major axis, approximately. As always, when we're talking about the sizes and distances of even relatively nearby galaxies, the watchword is "approximately," though astronomers are pinning down the true cosmic distances more closely every day. Like most elliptical galaxies, M59 is accompanied by an unbelievably huge retinue of globular star clusters, 2000+ in this instance. The Milky Way, in comparison, owns an unimpressive 150 known globs.

M58

Let's press on! When you tire of admiring M59's ghostly oval flying saucer shape, it's on to M58, a bright barred spiral (an SBc). Getting there requires only a small jump, as the galaxy is 1° 3′ west of M59. In the field with the galaxy is a noticeable magnitude 8 star. The galaxy itself has a magnitude of 10.8, so it is on the dim side. At least it's small at 5.0′ × 3.8′, so telescopes of 8 inches and above in aperture should make quick work of it. I have noted that on poorer nights it can range from "invisible," to "almost there," to "hard" with my 8-inch $f/5$. Boosting the magnification to the 200× level always seems to bring it into view eventually. In a 10–12-inch scope, it's no problem whatsoever, even when it's hanging over the skyline in the early spring:

> I was hoping to see something other than another round smudge, and, at 200×, I believe I have. As the sky darkened to astronomical twilight, this galaxy grew to over 2′ in size in the 12.5-inch scope, and, in addition to a bright core and an outer envelope of nebulosity, I thought I could barely detect clumps or condensation in the nebulosity that might be signs of its spiral arms. At lower power, this is just another featureless Virgo ghost.

If you hope to see anything beyond the smudge level, be prepared to spend a lot of time and effort on M58. Out in the dark country, 8-inch telescopes have a chance of seeing indications of this galaxy's central bar, but this feature is considerably easier in larger telescopes, in which the galaxy begins to take on a spindle shape. You'd never know it to look at it, but this is one of the brightest galaxies in the Virgo Cluster. Our ability to see this large galaxy in the city is helped greatly by its bright nucleus. While not as obviously active as the nuclei of "peculiar," disturbed Seyfert galaxies like M94, M58's nucleus' high-luminosity does indicate that something similar may be going on in this galaxy.

M89

When you're ready, jump out into space again, 49′ to the northwest. The next field contains the magnitude 9.8 E0 elliptical M89. Like most smaller ellipticals (it's 5′ × 4.6′ in size), M89 shows up very well and is no challenge at all. In addition to the glowing ember that's M89, there are a few stars of 8th to 9th magnitude in the field, and quite a few dimmer ones, making for an attractive vista in my 12.5-inch Dobsonian. In that telescope, the galaxy looked good, if somewhat bland in typical elliptical fashion:

> Very bright with a stellar nucleus. A round featureless envelope of nebulosity surrounds this center, extending out about 1′ tonight. Bumped the magnification up to over 200×, but no additional detail beyond a little more nebulosity is seen—not that I expected much more than a round fuzzball from an E0.

Everywhere you go in Virgo, you find "Es." The presence of so many elliptical type galaxies may be due to the generally high density of galaxies in the Virgo cluster.

These plentiful ellipticals may be the final result of the collisions and interactions that must occur continuously within central Virgo. Like many elliptical galaxies, M89 is a radio source, and possesses a jet of matter flowing out from its nucleus and extending over 100,000 light years into space. This activity may be a sign that M89 is suffering "indigestion" after having recently swallowed a smaller companion.

M90

Once you're on M89, M90 is close at hand. Sweep—slowly, very slowly—40' northeast and you'll be on M90's home field. Now, it is a good time to talk about identifying fields and landing on the *right* galaxies. When you've got two objects as close together as M89 and M90, a mere 40' apart, a situation that you'll run into constantly in Virgo, there's always the chance of misidentification. Work slowly when you move onto your next target, looking intently at everything that passes through your field. Once you think you are on the correct object, use your finder chart to try to identify patterns of field stars.

There are almost always enough stars in any given area to allow you to positively identify your target. To make this as easy as possible, print out a whole series of charts with only one galaxy per page, showing stars right down to the limit of your scope's ability. Use these in conjunction with your "main" eyepiece hopping charts to help ensure you're on the "right" galaxy. Another help in positively identifying objects is remaining constantly aware of directions in your eyepiece and on your chart. If you've got two objects in the field of your eyepiece, reference your chart and determine what their compass positions should be relative to each other.

How do you tell if you're *really* on M90 and not M89? That's easy—if you know which direction is which in your eyepiece field. M90 is the more *northerly* of the two. Another easy tip-off is the way this one looks. It is nothing like the E0 type M89. M90 is an Sb spiral, and it *looks it*, showing obvious elongation, even with modest telescopes.

What did M90 look like once I was sure I was on it? It is large at 9.5' × 4.5', but even at magnitude 9.5, it, like the rest of the Messiers in Virgo, is not usually a challenge for a medium-aperture telescope:

> M90 is a large elongated galaxy in photos, but tonight, with my 12.5, it appears consider-
> ably smaller than either M58 or M89, and barely subtends 1'. Still, though, its oval shape
> is detectable, if not as easy as I've seen it in the past. Small, bright, prominent nucleus. By
> any standards, this is a dim DSO tonight. Don't know that I'd have seen it in an 8-inch
> scope.

In long-exposure photographs, M90 is much more interesting than it is visually in the hazy air and light pollution of my backyard, showing off a prominent central bar and several dusty patches in its images. This is one of the bigger spiral galaxies in Virgo, but it's not very massive. It also seems that this is a very sedate galaxy despite its Virgo cluster address, showing no signs of an active nucleus and little or no evident star formation in the spiral arms. These characteristics lead some astronomers to wonder if this might be a "transitional" form between spiral galaxies and the weird lenticulars, which, like ellipticals, have no star formation going on. Maybe. In most long-exposure photographs, plenty of dust is visible across the galaxy's disk, something ellipticals and lenticulars lack.

M87

Probably the easiest way to get to M87 is to *back up*. Instead of making a big jump to The Monster Galaxy from M90, go *back* to M89. From there, it's a straight, if still sizable leap in the dark to M87. Move 1° 12′, two medium-wide eyepiece fields, as shown in the chart in Figure 6.9. On all but the brightest and haziest nights there should be no doubt when you get to M87. It's a big, bright elliptical at magnitude 8.6 and 7′ in diameter. Like the rest of its elliptical kin, its large size doesn't make it difficult to find—it stands out like a neon sign:

> This monstrous galaxy is beautifully bright in the C11 SCT. Looks just like M13 does in my 80-mm refractor at low power—a round, unresolved, and smooth ball of light. There's a bonus object here, too, the faint smudge of galaxy NGC 4476 12.5′ from M87. This small spiral galaxy is just a fuzzy mote, but is easy to see. A *Skytools* chart shows four other small galaxies brighter than magnitude 14 in this field, but I don't see any of them. As for M87, I don't see a nucleus, just a smooth brightening to its center.

I keep referring to M87 (Plate 33) as a "monster" or "monstrous." Why? This thing is *bizarre* and is of a size and majesty that's difficult for the human mind to encompass. M87 stands alone and apart from its 2000 fellow residents of the Virgo cluster. It's the *biggest, brightest* (intrinsically), and *strangest* of them all. At least 100 times the mass of our not-exactly-small Milky Way, and blazing with the fire of a *trillion* Suns, M87, sits near the center of the Virgo cluster, a fat old spider in its web, gobbling any of the unwary inhabitants that pass too close to its massive form. And it sings. It warbles a mindless song across the dark light years in its guise of radio source *Virgo A*. M87's radio voice is the result of violent processes taking place in its nucleus. In addition, M87 spits fire across 5,000 light years with a jet of tortured matter that is so luminous that it has even been detected by very large amateur telescopes from very dark skies at very high magnifications. The only engine we know of that can generate these titanic energies is a truly massive black hole.

M84 and M86

M84 and M86 are both classified as lenticular S0 galaxies, lens-shaped featureless bodies that don't reveal any details beyond their basic shapes, no matter how big you go telescope-wise or how dark you go sky-wise. The wondrous thing here is that you've got two bright galaxies (magnitude 9.1 and 6.7′ × 6.0′ in size, and magnitude 8.9 and 9.8′ × 6.3′ across, respectively) only 16′ apart.

These two galaxies, shining like bright cat's eyes in the darkness—the pair has long been known to amateurs as "The Eyes"—are also, amazingly and amusingly, accompanied by a "nose" and a "mouth." The nose is formed by small, round magnitude 12.0 NGC 4387, and the mouth is made by the appropriately edge-on magnitude 11.0 NGC 4388. When you've got the conditions and aperture required to see the fairly challenging nose and mouth in addition to the easy eyes, these distant, wondrous objects taken on the positively comical appearance of a 1970s "have a nice day" happy face (Plate 34). This is a great field even if *all* you can see is The Eyes, however (sketched in Figure 6.10):

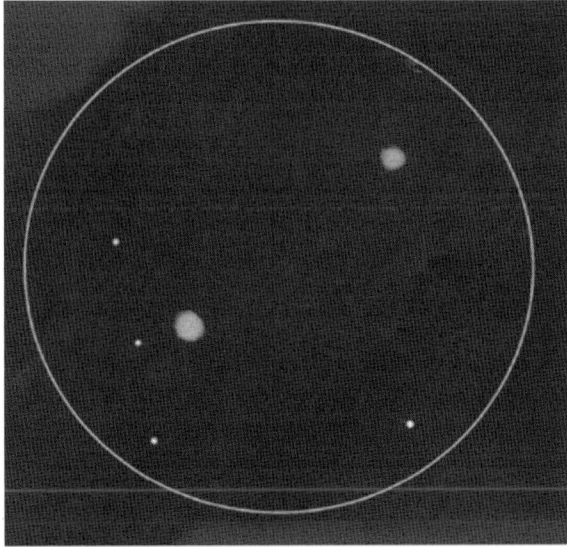

Figure 6.10. M84 and M86, The Eyes.

The eyes both appear round, with M84 being the larger and brighter of the two. M84 also displays a tiny, obvious core—M86 does not. Both are beautifully framed in the field of the 12-mm Nagler in my 12.5-inch Dobsonian Newtonian. Tonight I'm only seeing the two bright galaxies in this usually packed field—Nose and Mouth are invisible.

Tonight's Double Star: Gamma Virginis , Porrima

Just as there are more and less challenging galaxies, there are more and less challenging double stars. Magnitude 3.5 Porrima is certainly bright and easy to find—it marks the spot where Virgo's Y shaped arms join her body—but it can be a little tough to resolve. The primary and secondary are both the same magnitude, 3.5, and that usually makes for an easy split—the toughest doubles are those with a primary and secondary star of very unequal magnitudes. However, these two stars are currently (1995) near their minimum separation. Their 169-year mutual orbit has brought them to within a difficult 0.4″ of each other, meaning that a 10-inch scope at high power is required to give even a hint of resolution.

Even when the pair was 3.5″ apart years ago, Gamma Vir was usually not quite split by my 4.25-inch scope, appearing like a little "figure 8" with the two yellowish-white stars' airy disks in contact. Despite this difficulty, Porrima is a great double star to observe as a long-term project, especially if you own a medium-aperture instrument.

It actually *does something*. Over the next several years, the separation between these two gems will begin to increase and the change will be detectable visually.

> "Packed field" is right. M84 and M86 reside very near the core of the Virgo cluster, and there are galaxies everywhere. Under a suitably dark country sky, this is one of the most rewarding spots in the heavens. In a 12.5-inch telescope from magnitude 6 naked-eye conditions, I can count at least seven other galaxies in this half-degree field in addition to the two lenticulars. This vista—or even just the two bright galaxies—evokes genuine awe. The combined light of how many trillions of suns is visible at once in my eyepiece? It makes me feel small and pitiful, and makes all the works of man seem insignificant. Then I take heart. The stars blaze on unknowingly, but we are able to take-in their majesty and use it to fuel our dreams.

Special Bonus Object
The Inhumanly Distant 3C 273

Cosmic distances are not easily grasped by the human mind. Even the paltry 1.8 billion miles from Earth to the planet Pluto is hard to come to grips with. Get beyond the home galaxy, out in the deep waters of intergalactic space, and the stated distances to the island universes lurking there become essentially meaningless. What's the difference between 20 million light years and 60 million light years when 2 billion miles seems incomprehensible? But the next step on the cosmic distance ladder, to the mysterious quasars, helps put everything in perspective. 60 million light years doesn't seem far away at all and seems easy to visualize when you contemplate an object nearly 3 *billion* light years distant.

3C 273, the 273rd object in the Cambridge catalog of quasars, "Quasi Stellar Radio Sources," is 2.6 billion light years away and is probably the most distant object you and your telescope can hope to view from the city. Surprisingly, considering how far away it is, it is trivial to see, shining at magnitude 12.8. A magnitude figure dimmer than 12 may make you a bit skittish after finding out that 10th magnitude galaxies are sometimes difficult to see with urban scopes, but the quasar is a star-like point, so it is not nearly as hard as a 12th magnitude extended object. 3C 273 is doable with an urban 6-inch telescope at high magnification, and is quite easy with 8-inch and larger instruments.

No, seeing is not the problem with 3C 273. It's the brightest of these odd objects. Unfortunately, finding and identifying it *are* difficult at best. The easiest way to locate this quasar is with a go-to-equipped telescope. Even then, be prepared to use a detailed finder chart generated with a computer program to help you to decide which "star" in your field is really 3C 273. If your go-to telescope has a "precision pointing" mode, this is the time to use it. The closer you can put the quasar to the center of your field after a go-to slew, the easier identification will be. Digital setting circle users certainly have an advantage over star hoppers, but even with DSCs you'll have to do considerable searching, as it's unlikely that your circles are good enough to put objects dead center in the field every time.

If you have neither go-to nor digital-setting circles, be prepared to work hard for this object. Even experienced observers should be prepared to spend a half hour to an hour looking for it the first time. Your starting position is within the arms of Virgo, to the east of the galaxies we've visited on this tour. The quasar forms a right angle with Eta and Gamma Virginis, being about 3° 30′ north of Eta and 4° 45′ west of Gamma. There are no bright stars in the area, so the way to find 3C 273 will be to work slowly with a series of detailed charts, matching the field stars shown on them to what you see in your finder and main scope.

Once found, what exactly will you see? Not a whole lot. In any scope. Even the Hubble Space Telescope has trouble picking out details in quasars. You'll see nothing more than a somewhat distinctive looking blue star-like point. It's not much to look at, no, but I hope you'll be content with the wonder evoked by the significance of 3C 273. You're seeing an object so old that when light set out from it, life, simple life, was barely establishing a foothold on our home planet.

Quasars were discovered in the late 1950s when astronomers began noticing odd radio sources. These were soon linked to dim star-like objects. When these "Quasi Stellar" objects were observed with spectroscopes, their spectra revealed Fraunhofer lines hugely shifted toward the red end of the spectrum, indicating the quasars are receding at huge velocities. This huge Doppler shift also means they are very distant, some of them near the edge of the observable universe.

Over the years, the best guess as to what quasars are has been that they are the visible signs of extremely massive black holes at the centers of young galaxies. These black holes must be feeding off torrents of in-falling matter, and are almost unimaginably bright. Quasars, then, are the ancient and far more active cousins of the active galactic nuclei seen in familiar objects like galaxy M94. The Hubble Space Telescope has revealed what appears to the haze of "host" galaxies around some QUASARS, seeming to prove this theory to be correct, but there is still much that is unknown about these ancient objects.

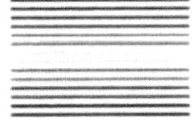

CHAPTER SEVEN

Summer

Tour 1

The Friendly Stars

One summer, I found myself living alone in the urban home of some friends, "house-sitting" for them while they were on vacation. I was recently divorced and hadn't yet readjusted to the single life. The empty house seemed strange and spooky, and the nights promised to be long. Luckily, I'd brought my 6-inch Dobsonian telescope with me, and when the skies cooperated I had no lack of friends to visit on these summer evenings. Looking out at the darkening sky on one surprisingly clear if substantially light-polluted evening, my gaze ran across the southeastern horizon and I found myself drawn to one of these friends, mysterious Ophiuchus, the Serpent Bearer and First Physician (in the guise of Aesclepius). His stars, glowing with burnished majesty in the still summer air, beckoned and I was soon setting up the telescope.

It had been some time since I'd visited this most-unfamiliar of the summer constellations (for nonastronomers, anyway) and I was anxious to see how his multitudinous globular star clusters would hold up in sodium-pink city skies. Sure, I'd visit Hercules—how can you pass up M13 on any evening when it's over the horizon? But, as I've said before, desert-only does not a nutritious meal make. My main target area for the evening would be Ophiuchus, with a side trip into neighboring Serpens Caput for a look at a star cluster I think is actually *better* than M13.

Let's get started. Before you can begin to sample the deep sky pleasures of Ophiuchus, you must to be able to find the constellation. That may not be easy if you are a new

Figure 7.1. The Ophiuchus neighborhood.

visitor to this out of the way part of the summer sky. Ophiuchus is composed of subdued stars that are spread across a large expanse of sky in a somewhat shapeless pattern. He's big and dim and it's difficult to trace his form in bright and humid urban skies, but to become a productive explorer of the summertime sky you need to extend your knowledge beyond well-known constellations like Hercules, Scorpius, and Sagittarius and start poking around in out of-the-way corners. You must stray off the beaten path and wander among the strange constellations you never hear mentioned on *Star Trek*, constellations like Scutum, Lupus, and *Ophiuchus*.

Many of Ophiuchus' stars are readily visible under the streetlights, but, no, his form does not jump out at you as that of nearby Hercules does—and Hercules, despite his fame, is not blessed with any bright stars. The basic pattern of the constellation is that of a house shape, a square with a triangle attached (see Figure 7.1). Ophiuchus' "house" is lying on its side, and the peak of the slightly distorted "roof" is pointing north. Don't even try to visualize the pattern as a human figure holding a snake Ophiuchus looks no more like a physician than Sagittarius looks like an archer or a centaur.

The easiest way for the new explorer to land in Ophiuchus is to follow the stars of Hercules. Start from the Keystone, the central pattern of stars that forms the Hero's body, move up his Eastern "arm" from the Keystone to Delta Herculis, and from there move to Alpha Herculis. Exactly 5° 15′ almost due east of Alpha you'll find one of Ophiuchus' few memorable stars, lustrous magnitude 2.08 Alpha Ophiuchi (Rasalhague), which forms the peak of Ophiuchus roof.

When you've identified the constellation and can pick out its shape with ease, look for the star Delta Ophiuchi (Yed Prior), in the southwestern area of the constellation figure—it forms the southwest corner of the house. Once you've got this magnitude 2.74 sun in the in your finder, look just 1° 24′ East for Epsilon. Pass this bright one by and keep going for another 7° 30′ and you'll happen on magnitude 2.56 Zeta Ophiuchi. Zeta and Delta, along with a dimmer star, magnitude 3.82 Lambda Ophiuchi, Marfik, which lies to Delta's north, will make it easy to find our first target, M12. Move your scope slowly and methodically, though, as there are no overly bright stars in the immediate area of the star cluster to help you find your way.

M12

M12, which lies 1° 39′ southeast of Lambda Ophiuchi, forms a near-right triangle with Lambda and Delta. If you position your scope with reasonable care, I wouldn't be surprised if you hit this cluster on the first try. If you miss it, a quick search should be all it takes, as, at magnitude 6.1, this Shapley–Sawyer Class IX (9) cluster stands out prominently, even in the bright and humid skies of city summertime. Its classification as a "9" means it is fairly loosely concentrated, without a prominent core, but it's far easier to run down than most of the galaxies of spring were.

What does M12 look like in a 6-inch aperture telescope? It can be disappointing. You'd think the "loose" nature of this globular would mean even a small telescope would deliver lots of resolution; cluster stars shouldn't be so closely packed together as to prevent a 4 or 6-inch from plucking individuals out of the milky glow. Unfortunately, this loose structure actually makes M12 a difficult subject for a small urban scope under gray summer skies if your goal is the resolution of individual cluster stars. Far from spectacular, M12 reminds me of a slightly brighter twin of spring's M53. Yes, M12's stars are well separated, but they are dim. While a few of this glob's suns approach magnitude 12, most are at a forbidding magnitude 14 or dimmer. The cluster's light is spread out across a fairly large area of sky, too, 16′, making it even less impressive under poor conditions. As I found out during my house-sitting observing runs and documented in Figure 7.2 and in a log entry:

> I had great hopes for M12 based on what it looks like from the country, where it's a spectacular (if not overly bright) ball of stars in the 6-inch Newtonian. But it's not nearly as good from this urban site as I'd expected. It's a vague and ghostly glow 3′ or 4′ in size at most in my scope. There is only a little central condensation. Mainly it's just a small gray disk that dims smoothly to its edges. Despite increasing my magnification to 250×, and using averted vision on the cluster's periphery, I can't resolve any stars tonight. Most of the Messier globulars will give up a star or two if only briefly and with averted vision, but not this one.

Keep your eye on this cluster throughout the summer, of course, as you never can tell when a night of improved observing conditions will make it look *radically* better in smaller telescopes. The easiest cure for the can't-resolve-the-glob blues, though, is to pour-on the aperture, as I did recently with my Nexstar11 SCT:

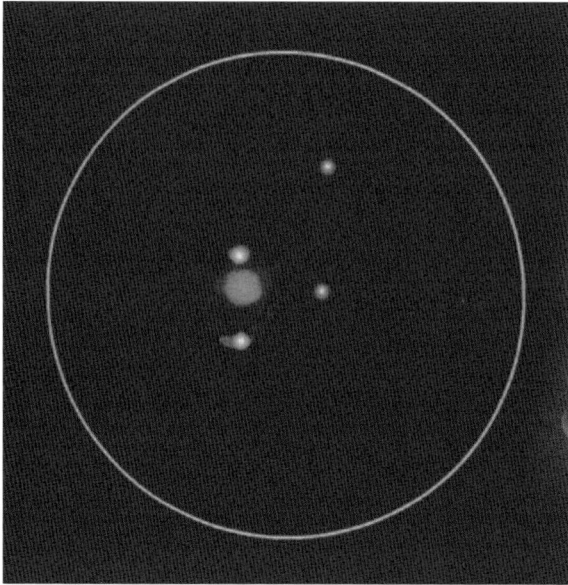

Figure 7.2. M12, one of the twin globulars of Ophiuchus.

M12 is excellent despite the haze-amplified light pollution I'm dealing with tonight. Much resolution at both 220× and 127×. The loose core is not completely resolved, but numerous stars, both around the edges and closer in toward the center, are easily visible with direct vision. Looks vaguely irregular, with an almost "square" appearance. It's large, and I wish I could use my 35-mm eyepiece on this one, but it makes the background sky way too bright.

M12, like the constellation Sagitta's famous M71, is so loose in structure that astronomers have wondered whether it might really be an open cluster rather than a globular. Spectroscopic studies, however, reveal its stars as elderly rather than youthful— open cluster stars are always young compared to those that compose globulars. M12, which lies an estimated 16,000 light years from our planet, is an ancient globular star cluster. If it is indeed 16,000 light years away, it is about 75 light years in diameter. Like other globs, it is composed of old stars, both evolved ones that are in the red giant stages of their lives, and small, low mass stars that can continue to shine dimly for countless eons.

M10

There are two ways to find the nearby (to M12) globular, M10. If you're an experienced observer comfortable with your scope, just eyepiece-hop 3° 16′ east from

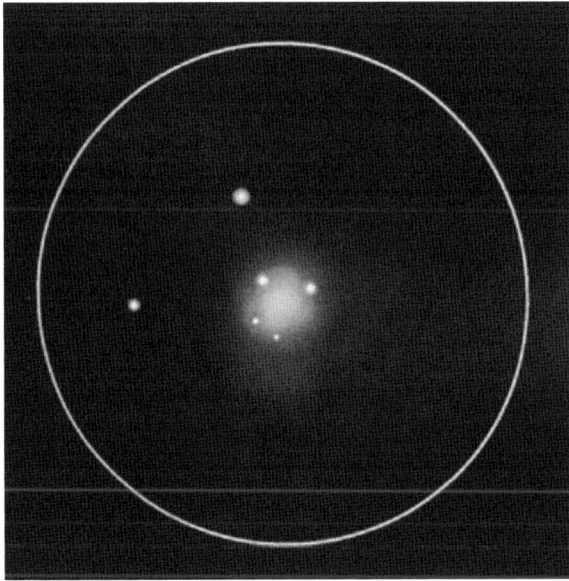

Figure 7.3. M10, Ophiuchus' other twin.

the previous object, M12. If you're *not* so experienced, or, like me, don't visit Ophiuchus as often as you should, use the stars Delta and Lambda again, but also include nearby Zeta Ophiuchi in your pattern making. As seen in Figure 7.1, M10 forms an almost perfect rectangle with these four stars. If you have trouble hitting the right spot this way, look for magnitude 4.8 30 Ophiuchi in the area of the cluster. It shows up well in finderscopes, and M10 is only about a degree away from this star. At magnitude 6.6, M10 is a little dimmer than M12, but this is more than made up for by its Shapley–Sawyer Class of VII (7). Its more compact form gives it a brighter central region, making it stand out better from the gray sky background in your eyepiece.

Since M12 and M10 are so close together in the sky, it's inevitable that they'll be lumped together as "two of a kind." You'll often hear them referred to as the "Twin Globulars of Ophiuchus." But that's ridiculous. They *do* have similar sizes and magnitudes, but they look *nothing* alike in an eyepiece. M10 (Figure 7.3) is *worlds* better than M12 from the city. Or in the country, if you ask me. M10 is bigger and dimmer than M12, and logically should look worse, but its more compressed structure makes a huge difference. In the 6-inch scope I found that:

> M10 is much more impressive than M12. In the 6-inch scope at 150×, even with the cluster lower in the sky, numerous member stars are resolved around its periphery and can be held steady in my gaze with averted vision and high power. A few even wink in and out of view when I stare straight at the cluster.

Naturally, going to a bigger scope makes the good even better:

> Lots and lots of resolution with the 12.5-inch Dobsonian at 125×. I can see cluster stars
> right across the globular's core with direct vision. This core almost looks "transparent"—
> as if I'm seeing through it, past its multitudes of stars to empty space beyond. Streams
> of stars branch off from the cluster's center and almost seem to assume a spiral pattern.

It doesn't take a 12.5-inch telescope to make M10 look good. This is a truly remarkable object in my inexpensive 8-inch *f*/5 equatorial Newtonian as well. At 167× with
an imported 6-mm wide field eyepiece, M10 just looks *amazingly* wonderful. In the
heaviest light pollution, many small stars spill across the cluster core and occupy almost the entire width of the eyepiece's 60° apparent field. I believe I actually preferred
the view in the 8-inch scope to that in the larger Dobsonian. The 8 inch's driven mount
made it easier to study the cluster and to see all the stars it could deliver—continually
nudging the Dobsonian along to track the cluster was distracting. The 8-inch *f*/5 was a
significant improvement over the 6-inch scope for very little penalty portability-wise.

Unlike M12, there has never been any doubt about the nature of M10 (Plate 35). A
glob is a glob is a glob in this case—the aged stars revealed in its HR diagram and its
appearance in any telescope make that clear. It's a little closer to us than M12 at 14,300
light years, which makes it a little smaller than its brother at 63 light years in diameter.

M5

I've gotten into *arguments* about this Messier globular cluster. If you live in the Northern Hemisphere, all you hear is, "M13, M13, M13." Hercules' Great Globular Cluster,
is good. OK, *very* good. But is it the best globular cluster for observers at northern latitudes? Not in my opinion. There's, for example, M22 in Sagittarius. It's much bigger,
looser, and easier to resolve than M13 in small scopes. Too bad it's so low in the sky
for many Northern Hemisphere observers. But there's another globular, one that, to
my eyes, looks considerably better than M22 *or* M13, M5 in Serpens Caput (Plate 36).

If you thought Ophiuchus was an obscure constellation, I'm betting Serpens is
completely unknown territory for you. Serpens is actually two constellations: *Caput*
and *Cauda*. As the Latin words indicate, one is the Head of the Snake and one the Tail.
Visualize the figure of Ophiuchus/Aesclepius with his familiar snake draped across
his back, head on one shoulder and tail on the other. The star-pattern of Ophiuchus
doesn't look much like a man, but the constellation figure of Serpens Caput does look
something like a serpent's head. We can forget about the Cauda, the Tail, for now, as
M5 is within the borders of Caput.

Serpens Caput is easy to find if you know the basic form of Ophiuchus. It's connected
to the main figure of the constellation at the star Delta Ophiuchi, which we used to find
M10 and M12, and extends through four medium bright stars that lead you to Serpens
Caput's most distinguishing feature, the head of the snake formed by Rho, Iota, Beta,
Gamma, and Kappa Serpentis. While it might be argued that Serpens looks more like
a snake than Ophiuchus looks like person, forget that this asterism is supposed to
represent a snake's head; the five stars form a nice letter "X" in the sky southwest of
Ophiuchus. Always look for this X, as it's usually quicker to locate Serpens by means of
this prominent pattern than by wending your way up from the east from Ophiuchus.

With your eyes fastened on the "X," note magnitude 3.67 Beta Serpentis, the southeastern member of the asterism. From there, jump down to the next prominent star along the snake's back, mag 3.80 Delta. The snake's body jogs back North after Delta, and you'll next run into Alpha (magnitude 2.65), and Epsilon (magnitude 3.71) fairly close to each other. M5 is 9° 21′ Southeast of Delta, Alpha, and Epsilon, forming the apex of a slightly lopsided triangle in combination with their slightly tilted base. Another way to find the cluster is by using magnitude 3.88 Mu Virginis, which is 11° 46′ to the southwest. M5 is a little over halfway along a line drawn from Mu to Alpha Serpentis, and is a little closer to Alpha than Mu. M5 is not difficult. At magnitude 5.6, you can safely use a low-power wide-angle eyepiece while hunting for it. This star cluster burns through light-pollution with ease despite its large size of 22′.

It didn't take me long to center M5 in the field of the 6-inch scope, even using my way-too-small 30-mm finder. When I went to the eyepiece, I liked what I saw:

> Lovely, with a very compact core. It appears slightly elliptical, and is elongated east/west. Raising my magnification to 200× reveals plenty of outlying stars away from the tightly packed nucleus—which doesn't show any resolution in the 6-inch scope. Other than its elliptical shape, there's no real form or pattern to its stars. The stars surrounding its core congregate in random patterns, nearly filling the field of my eyepiece. I hate to leave M5, but there's so much else to wonder at in Ophiuchus and the rest of the summer constellations—I've got to move on.

Like spring's M3, M5 often looks blue in color to me. You'd think a globular, made up of old yellow and red stars, ought to have an overall golden hue, but this one glows an amazing sapphire. While I used a 6-inch scope to view M5's minute stars, they should be visible in 4-inch telescopes at high power, since the brightest cluster members shine at magnitude 12.

M5, a Class V (5) globular, is one of those Messier objects that Messier himself didn't originally discover. It was first seen in 1702 by German comet hunter Gottfried Kirch and his wife Maria during one of their "sweeps" for comets. Charles Messier did discover it independently in 1764, and recorded it as a "nebulous star." M5 is one of seven globulars on Messier's list that lie inside the borders of Ophiuchus. This enormous ball of suns, 126,000 light years away from our cozy rock, is incredibly ancient. It is believed to be one of the most elderly of the Milky Way's globulars, old enough that it may have been witness to the titanic events surrounding the birth of our galaxy.

Other Globs and a Sweet Surprise
M107

Lurking near the center of the galaxy, which is found in the nearby constellation Sagittarius, the Ophiuchus area claims more than its fair share of globulars, most of which are at least *visible* in small telescopes. If you want to see more than the three showpieces in this region, M10, M12, and M5—and you should—M107 is a good place to start, as it's even easier to find than M10 or M12. It's dimmer and smaller than the other two at magnitude 7.9 and 10′ in diameter (in pictures—you'll be lucky to see less than half that much), but its location near prominent Zeta Ophiuchi, means that finding it is simplicity itself. Position your finder on Zeta and slew 2° 43' south. Just before you hit the cluster, you'll cross a lopsided triangle of 7th magnitude stars half a degree across. Since M107 is going to be small, kick up your magnification as much as you dare. Even though it's very loose, M107 is small and high power will help you pick it out from the background.

What M107 will look like in your telescope depends, as it does with every other DSO, on you, your skies and your scope, but it was at least worthy of a few minutes with my 6-inch:

> Surprisingly dim in the 6-inch Dobsonian, even at 150×. I can see it, but can't resolve a single star. It's fairly identifiable (at 96×), nevertheless, and I believe a better evening and higher power will improve it.

Coming back to it later with the C11, it was better, certainly, but admittedly the bigger telescope still didn't make it spectacular on a poor night:

> M107 is there this evening—barely. At 220× with direct vision I can make it out with some difficulty. Even though it's fairly high in the sky, it just doesn't look like much. It's a small, gray, fuzzy ball with only occasional resolution of a few stars around its edges at 300× and higher magnifications. At lower powers, it is completely unresolved. I'm gratified to be able to see any stars at all in this small cluster.

NGC 6235

Want *more* of a challenge? A globular cluster that's not a Messier? Trot over to the far southeastern corner of Ophiuchus for NGC 6235, another very loose cluster with a Shapley–Sawyer Class of X (10). Not only is it loose, it's dim at magnitude 10.2, and small at 5'.

NGC 6235 lies almost exactly at the midpoint of an imaginary line drawn between two easy to see stars, magnitude 2.43 Eta Ophiuchi, and magnitude 2.82 Tau Scorpii, which is 14° 49' south of Eta. This position close to two prominent stars would ordinarily make the cluster easy to find, but this globular is faint, so keep your eyes open. As you move south toward Tau, you'll run across magnitude 6.26 29 Ophiuchi, the only marginally luminous star close to the cluster. It is almost exactly halfway between Eta Ophiuchi and the glob. Just before you reach the star cluster, you'll see two 7th magnitude stars aligned almost perfectly with the line you've "drawn" from Eta. They are approximately 22' apart, and the second star is only 25' from the NGC 6235.

Even in the country, this distant cluster doesn't give up its secrets easily. It may take a 20-inch class telescope to reveal a handful of stars under dark skies. Knowing this to be a very real challenge, I didn't even try for it with the 6-inch scope during my house-sitting stint—I came back to it later with the 12.5-inch Dobsonian:

> NGC 6235 is low on the horizon, skimming the trees, and in some of the worst horizon garbage. I wasn't sure whether I'd be able to see this cluster at all, even with the 12.5-inch scope. It is detectable, but only barely, as a small, 1' or so diameter smudge at 127×. Can't always hold it with direct vision; it pops in and out. It looks more like a dim Virgo galaxy than a globular cluster. Higher magnifications don't help either; 220× makes it disappear completely.

M62

M62 is a much more practical cluster for small scopes and bright skies than NGC 6235. It's at magnitude 6.5 and is a Class IV (4), so it is rarely a challenge to see. The only challenge, in fact, is trying to achieve resolution in smaller telescopes, but even in that regard M62 is a kindly and forgiving old cluster.

M62 is a little over 7° from NGC 6235, and is more a part of Scorpius, really, than Ophiuchus. It straddles the Ophichus—Scorpius border and is less than 4° from Epsilon Scorpii, one of the Scorpion's "backbone" stars. It forms a near isosceles triangle with Epsilon and Tau Scorpii. Both of these stars are very bright, but proceed slowly. Unless your observing site is close to the equator, you'll be working down near the horizon, and the stars and the cluster will be dimmer than you think they should be, even in a large aperture finder. Once you've got the cluster in the field of the main telescope, pour on all the power that it, your scope and your skies will support if you want to pick out a few of this glob's distant stars. I worked hard on this one, spending over an hour with it and using a wide variety of magnifications with my 6-inch *f*/8. This diligence rewarded me with at least a brief glimpse of its true form:

> This is more than just another faint fuzzy. At 180×, averted vision picks up a few cluster stars every now and then. To achieve this modest triumph, though, I had to cover my head with a dark hood and spend a lot of time looking. In addition to averted vision, I found that lightly tapping on the telescope tube, making the image of the cluster vibrate slightly, was necessary to make the few visible stars easy to see.

NGC 6572

I mentioned a surprise, didn't I? In addition to globular clusters, Ophiuchus, as you'd expect from his close proximity to the Milky Way, holds many of those other typical galactic-plane type objects: nebulae, open clusters, and planetary nebulae. A look at a detailed chart reveals them scattered all across his form. The Ophiuchus pages in a good star atlas look hopelessly jumbled from DSO symbols piled on top of DSO symbols. Despite the density of objects in this area, though, there's really only one other object of interest to the urban observer in addition to the prominent globs. But it's a wonder.

At a brightness of 9th magnitude, NGC 6572, "The Blue Racquetball Planetary," sounds discouragingly dim, but its tiny size of 11″ means this is no Owl Nebula. It's bright, standing out like the proverbial sore thumb in almost any telescope. To find it, head back to the northern part of Ophiuchus, in the direction of Hercules, and locate Alpha and Beta Ophiuchi. Moving Southeast from Beta, hit magnitude 3.75 Gamma and make a hard turn to the east and down to two fairly prominent magnitude 4.0 stars, 67 and 70 Ophiuchi. Now, turn north and go 5° 15′ and you'll encounter 71 Ophiuchi at magnitude 4.64. It's easy to spot since it has a companion, magnitude 4.0 72 Ophiuchi, just 49′ farther north. The nebula lies 1° south of the middle of a line drawn between 70 and 72. This sounds complicated, and I'll admit that a go-to telescope would be nice when hunting this sucker, but keep after him with a good atlas and you'll bag him—eventually!

Your biggest challenge won't actually be finding the nebula, but distinguishing the Blue Racquetball from a field star in a small telescope. At 11″ in diameter, this VV type 2a planetary is only one-fourth as large as Jupiter is at his largest. That means magnification, magnification, magnification. You should be able to identify a "star" in the field that looks slightly peculiar even at lower finding powers, however. Center this, run the power up to 200×, and you will get a nice surprise:

> No wonder they call this thing blue! If you want color in a DSO, this is the place to go. The Blue Racquetball is almost startling in the richness of its color. The blue of this planetary is more noticeable in the 6-inch scope than is the color of Andromeda's more famous Blue Snowball. The shape also lives up to its name. It is a featureless disk, just a little round racquetball.

If you mention your experiences with the Blue Racquetball at an astronomy club gathering, you'll probably get a lot of blank stares, as the word doesn't seem to have gotten out on this magnificent nebula. This is a tremendous DSO, so spread the word far and wide about this odd little ball bouncing through dim southern stars.

Tonight's Double Star: Marfak, Kappa Herculis

Marfak is an easy to split double star that shows off an interesting color contrast. With a separation of 28″, it is even practical for binocular observing. The magnitude 5.0 primary is a pure looking yellow, but the secondary, shining at magnitude 6.5, is best described as "bronze" or "copper." While Kappa is not part of the main Hercules stick figure, it's pretty easy to find. It lies a little more than halfway along a line drawn between Gamma Herculis, the "last" star in the Hercules, pattern before you reach the Serpens Caput border, and Gamma Serpentis, one of the stars of Caput's "X", which lies 7° to the west of Gamma Herculis.

> That's it for tonight's guided tour. Now, go your own way, wandering through the bewildering treasures of the south. That's what I did on this lonely summer's evening. I traveled from wonder to mystery to marvel as the night grew old and the dawn of another lovely summer morn crept up on me and my wonderful telescope. Lonely? By the end of the night I didn't feel a bit lonely. After soaking in so much beauty, the night and the universe didn't seem mysterious or forbidding at all. Even the distant, enigmatic globular clusters seemed no stranger than friendly bees buzzing around the hive of the Milky Way's center.

Tour 2

Arkenstone in the Stars

When darkness finally arrives after another long midsummer's day, amateur astronomers are witness to an incredible spectacle: the rising of the center of our home galaxy. Look to the south, and, in dark skies, you'll see the Milky Way's nucleus blooming like some monstrous fiery flower. The entire area of the southern Milky Way is literally packed with treasures for the deep sky observer. While the full beauty of this region is denied to the urban astronomer, there are still some lovely views to be had for even the smallest urban telescopes.

Of particular interest in the summer Milky Way is the area of an ancient zodiacal star figure, Sagittarius the Archer. This constellation, which represents the noble centaur Chiron, the teacher of Hercules of Greek myth, is prominent in the south even from fairly high northern latitudes. Don't look for archers or centaurs, though. Look for a heavenly tea service. As shown in Figure 7.4, the stars of Sagittarius form a perfect representation of a teapot.

What makes Sagittarius interesting for us deep sky trekkers is the fact that it is smack-dab in the middle of the summer Milky Way—the center of our home galaxy lies within Sagittarius' borders near M24, a rather subdued (in the city) open cluster/star cloud. This means the area is literally bursting with star clusters, both open and globular, and many wispy emission nebulae. The same goes for nearby Scorpius. It may not be quite as rich as the Archer, but it has some wonderful globular star clusters, and a few open clusters that beat anything of their type in Sagittarius. Scorpius, by the way, is one of the few constellations that beautifully portray what they're supposed to represent; in this case, the scorpion of myth that stung Orion to death.

The urban observer does have to be a bit discriminating when hunting objects in this region, since some types of DSOs show up better in light-polluted skies than others. Depending on how bad your site is, the southern nebulae may be practically exterminated by the sky glow. We can pull some of their beauty out of the light pollution with light-pollution reduction filters, magnification, and skill, but there is no denying that they are better seen from the country. Many of Sagittarius' open and globular star clusters, though, hold up well even from the worst sites.

For many of us, our options are further complicated by the nature of summertime weather. Like most observers, I have a love/hate relationship with the summer skies. On one hand, the coming of this season brings a large increase in the number of bright DSOs that are visible. On the other hand, crystal-clear skies are rarer now. Even the most beautiful days segue into hazy, muggy nights. These less than ideal skies become even worse when coupled with light pollution. Winter skies look darker whether you're in the city or the country because, as was mentioned in the first part of this book, they are *drier*. Less light is scattered by moisture in the air. If you live in an area with high summertime humidity, and have trouble with any of these objects, wait for one of those rare dry—or at least drier—summer nights and try your luck again.

Actually, it's unlikely any of tonight's DSOs will prove challenging, no matter how high your humidity level. All are bright. Amazingly bright. In fact, I had a difficult time

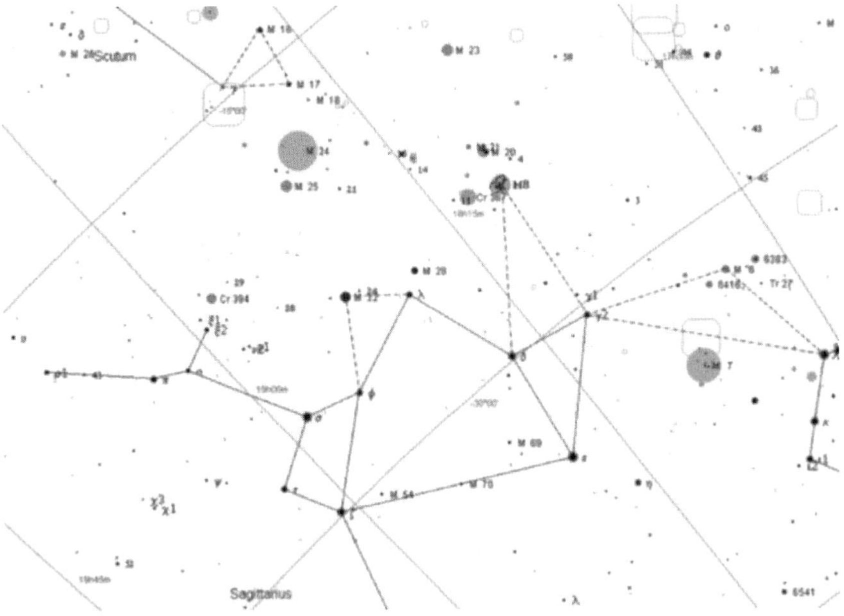

Figure 7.4. The Sagittarius Teapot.

putting this tour together. Not because of any lack of inspiration. My problem was that even when the less than optimum summer observing conditions of my backyard were taken into account I was still faced with the task of choosing a few objects from a multitude of beautiful sights. I really wrestled with which objects to keep and which to throw out, and finally came up with a selection that I think represents the very *essence* of the summer deep sky.

M80

Tonight's cosmic journey inward toward the center of the galaxy begins with the beautiful globular cluster M80 in Scorpius. Besides being a striking sight in fairly small telescopes, M80 also has the advantage of being simple to find. A quick look at an atlas locates this cluster 4° 27′ northwest of Scorpius' brightest star, the impressive red supergiant, Antares. An easy way to snare it is by using magnitude 2.89 Sigma Scorpii, which is just northwest of Antares. Draw a line from Sigma and through magnitude 4.55 Omicron Scorpii. Now, extend this line the same distance again and stop. M80 lies 1° 15′ to the southwest.

Once you have M80 centered in the field of your main scope, your first impressions will be *small* and *bright*. Less than 10′ in diameter and with a brightness of magnitude 7.2, M80 appears stellar in small scopes at low powers. Higher powers will definitely increase the details you can see in many globular clusters, and will usually resolve a few peripheral stars for medium-aperture instruments, but that's not the case with M80. Unfortunately for the city-bound small-scope observer, while higher powers will make M80 look more like a globular and less like a fuzzy star, they will not come close to allowing resolution of the cluster's elusive suns.

M80 is composed of dim stars, 14th magnitude and dimmer, and is highly concentrated, being rated a Shapley–Sawyer Class II (2). It can be hard for smaller instruments to resolve stars in loose globular clusters, but a little scope often has even more trouble picking stars out of the really tight ones. The combination of very small size and dim stars make it impossible to resolve M80 in the city with much less than an 8-inch telescope—and a 10-inch scope makes the task much easier.

This trouble with resolution, though, is offset in my mind by M80's incredible brightness and the beauty of the star field in which it is set. This area of the southern sky is filled with intertwining dark nebulae (William Herschel's famous "hole in the heavens" is nearby). And bright M80, shining like a beacon amid misty star clouds and obscuring nebulae makes for an unforgettable sight. An entry in my observing log for the 4.25-inch *f*/11 Newtonian records M80's appearance on a below average evening as:

> A nice cluster, though now near the horizon. Round, very concentrated, somewhat stellar, appearing at low power. No hint of resolution or graininess. It's a round, star-like object in a rich star field.

Nice, but, as always, I want *stars*. I enjoy observing globulars in their unresolved state in small telescopes and binoculars, but nothing beats seeing them break apart into myriad suns. With that goal in mind, I came back to M80 with the C11:

> At 220× in the 11-inch scope, my impression of M80 is "small and grainy" rather than the usual "small and bright" of smaller telescopes. It is really not much different from what I could see in the 4-inch scope, just brighter. I was able to make out a small number of stars around the edges, but even at 466×, as high a magnification as I usually dare to apply to this glob, M80 is difficult to resolve.

M80 is located nearly 30,000 light years from Earth. A note of historical interest is that the nova *T Scorpii* burst from invisibility to 7th magnitude amid M80's dim stars in 1860, changing the cluster's appearance radically, and briefly making it into the most observed globular star cluster in the sky.

M22

If you found the inability of a small telescope to resolve M80 into stars frustrating, our next destination, the awesome deep sky wonder M22, will more than make up for that frustration. As I mentioned when discussing M5 in the preceding Tour 1, M22 easily "beats" M13 in my opinion, and is especially rewarding for the urban observer with a smaller telescope. In fairly severe light pollution, a little scope—even a 3-inch scope—will resolve *some* of M22's stars at high magnification if you're far south enough in latitude to get the cluster out of the worst horizon-haze and light pollution. In this same 3 incher, M13 remains a fuzzy, if bright, *blob*. M22 really is the archetypal globular cluster for small telescope users.

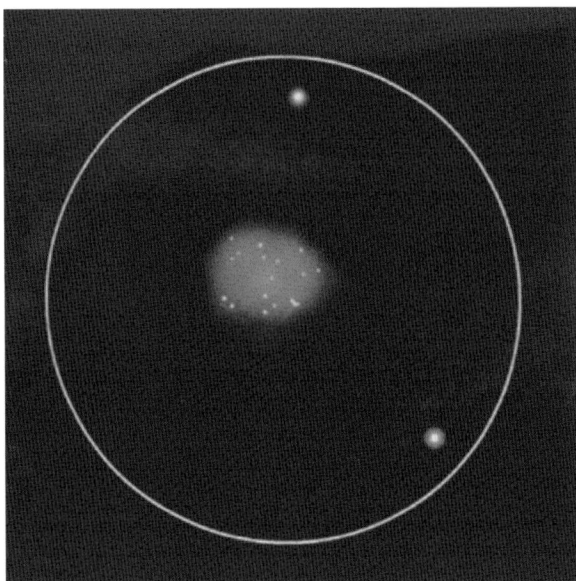

Figure 7.5. M22, Arkenstone in the stars.

M22 is also one of the easiest globulars to locate; it's 2.5° northeast of magnitude 2.8 Lambda Sagittarii, the "top" star in the teapot's lid. The cluster forms an isosceles triangle with Lambda and bright, magnitude 3.1 Phi Sagittarii, which marks the spot where "handle" joins the "pot" (Figure 7.4). Pop in a low-power eyepiece, move your scope into the general vicinity, and you should almost immediately locate the huge magnitude 5.1 glow that is M22. This thing extends a full 32′ across in photographs, making it slightly larger than the full Moon.

When you've got it in your sights, *gaze steadily at this "blob"* and you'll soon see *tiny* stars winking in and out of view around the edges, and will maybe even detect a few sprinkled across the cluster's center in a 4-inch telescope. A higher power eyepiece (100× in a 4-inch scope) should increase the number of cluster stars you can see. As always, diligent, prolonged observation of a DSO will reward the observer with many details that were invisible at first glance. In my modest Short Tube 80 refractor, this big thing was a delight, and I recorded its appearance in the drawing in Figure 7.5 as well in as in a text log entry:

Fantastic! This huge glob was getting awfully low by the time I finally pointed the ST 80 its way, but it still looked just, plain wonderful. M22 was bright in the field of a 15-mm TeleVue Plossl at 26×, but at that power there was not a hint of resolution. Boosting the magnification to a bit over 100× with a 7-mm Orthoscopic and a 2× barlow, however, did bring out at least a few cluster stars, and showed its strongly elliptical shape well.

In the C11, "good" became "oh-so-much-better":

M22 is low on the horizon on this average night and well into hazy, streetlight-pink skies, but it is simply incredible at 120×. It's a large, distended oval of tiny stars, and looks incredibly resolved. Very flattened. The core is a milky globe with many stars sprinkled across it. Averted vision brings out more there and on the periphery of the cluster. Nearly fills the field of a 22-mm Panoptic eyepiece. I don't notice any color in this glob; its overall hue is a silvery gray.

The technical details of M22 are as amazing as its appearance. Like most globular star clusters, it is mainly composed of Population II stars, the ancestors of the much younger Population I stars (like our Sun). This cluster contains, at a *conservative* estimate, at least 500,000 stars, many of which have gotten old enough to leave the main sequence, confirming this globular's advanced age. It is quite possible that M22 and its stars formed at about the same time as the Milky Way itself, 10–12 billion years ago. The diameter of M22's central region is 50 light years, and the cluster has been given a Shapley–Sawyer Classification of VII (7) (loosely concentrated). At an estimated distance of 10–12,000 light years, M22 is definitely one of the closest of these conglomerations of aged stars, being nearer to us than famed M13.

M17

Yes, light pollution takes quite a toll on the nebulae of the Southern Milky Way. Not only are these glowing clouds of hydrogen gas far more vulnerable to the ravages of sky glow than star clusters, the *good stuff* is down in Sagittarius and close to the horizon for almost all Northern Hemisphere observers. M17, the Swan or Omega Nebula, however, is always a pleasant surprise. At a medium size of $11' \times 11'$ and a magnitude of 6.0, even small instruments can reveal its presence from some pretty awful observing locations, and medium-aperture scopes can make it look almost as good as it does from the country if you can use an OIII or UHC filter to restore a little contrast.

The Swan is situated in the northwestern area of Sagittarius near the Serpens Cauda border, and is relatively easy to find. "Relatively" because, while this is obviously a star-rich area, there are no bright ones close to M17. The best path is probably from magnitude 4.7 Gamma Scuti. M17 is $2° 35'$ southwest of this star. There is a magnitude 5.3 star $23'$ west of the nebula if you need more guidance, but it's hard to miss M17 if you just move southwest from Gamma with a low-power eyepiece in the telescope.

Once you have M17 centered, you'll see why generations of observers have visualized this glowing cloud as a Swan, a Greek letter omega, or a Horseshoe, especially if you're using a 10–12-inch telescope and add a UHC or OIII to the eyepiece. The nebula (Plate 37) is composed of a bar that's $11'$ long in photos, and is terminated with a hook shape on its western end. Since there's not a bar on the western side of the hook, the nebula always looks more like a Swan than a horseshoe or omega to me. No matter which shape you see here (some people even see this as the *Lobster* Nebula), it looks marvelous:

> M17 is far better than nearby M8 or M20 in the C11 on this below average night. A rich field with the Swan shape floating among the many stars. Best seen at 100X with a UHC -type filter. I can't detect the dark lane that cuts through the bar of nebulosity that forms the Swan's body, but I can make out the swan's neck without trouble, though this crook of nebulosity is considerably shorter than it is from the country.

The Great Swan, which is a beautiful shade of pinkish-red in photos (it's gray to our less color-sensitive eyes), is a huge cloud of hydrogen 6,000 light years from us. If this distance is correct, it's 15 light years across—one big bird! It is estimated that the

Swan contains enough mass to make nearly 1000 Suns. M17 glows across the light years because its atoms have been "excited"—made to give off light—by the high-energy photons emitted by hot young stars hidden within its folds of nebulosity.

Interesting Side Trips in Sagittarius and Scorpius

M16

There are few nebulae more identifiable by the public than M16, the Star Queen or Eagle nebula. This is due to the famous Hubble Space Telescope image, the "Finger of God." In amateur scopes and images, however, it appears more like—though certainly not nearly as good—as it does in Plate 38. The Eagle is dim and difficult compared to the Swan, and even under dark country skies a UHC-filtered 8-inch telescope is *required* to easily reveal the nebula. In the city, seeing it is harder still, even with the aid of OIII and UHC filters. The nebula has a listed visual magnitude of 6, but, as with all extended objects, this is not a good indicator of how hard it is. In my experience, it's *far* more difficult than even its large 15′ size and surface brightness of magnitude 12 would indicate.

If you found M17, you shouldn't have any trouble getting the Eagle in your eyepiece. He's just 2° 27′ west, maybe two or three low-power eyepiece fields. Once you've got the bird centered in a medium-power eyepiece, don't be too disappointed if the nebula frustrates you. You'll at least be rewarded with the beautiful open cluster that has formed from the nebula's gas clouds:

> The famous Eagle is not an easy object in the C11. The rich open cluster is nice enough, of course, but even on this good night there is only the barest suggestion of the Eagle's nebulous body. Its "wings," which extend from the main body of the nebulosity, are invisible at high or low power and with all my nebula filters. The UHC seems to bring out what little nebulosity there is the best. I have never seen any trace of the Eagle with any scope smaller than 10 inches in the city.

M8

I always have high hopes for M8, the famous Lagoon Nebula. In the country, or even from decent suburban skies, this is a wonder. You have two patches of nebulosity cut by a dark lane, the "Lagoon", and a wondrous open cluster is superimposed on the nebula (this star cluster is not involved in the nebula, it is merely along our line of sight to it). Like M42, getting to M8 is like shooting fish in a barrel. Even under poor conditions, it should show up in your finder as a fuzzy star, 6° west of Delta and Gamma, Sagittarius' "spout stars."

Easy to find like M42, yes, but don't get the idea that this will *look* as good as Orion does in light pollution. M8 is badly hurt for us northerners by its closeness to the horizon. It's got a high integrated magnitude of 5.0, but this is spread out over an object 30′ × 45′ in size, so the surface brightness is rather low at 13.0. An OIII filter can help with the Lagoon, but there's no use mincing words: M8 is usually a huge disappointment if you're familiar with its appearance from dark sites:

> With a UHC or OIII filter only a small—maybe 10' across—patch of nebulosity is visible, in the 8-inch f/5, and that patch is faint. Other than this nebulous knot, which is clumped around the bright star 9 Sagitarii, the main attraction here is the fine open cluster, NGC 6530 just to the East of 9. Averted vision did seem to show up a little extra nebulosity, and I thought I occasionally caught hints of the dark lane that splits this object in two, but the Lagoon is a subdued and difficult object in the city.

M4

While not exactly easy in the city, either, globular cluster M4, back over in Scorpius, is certainly easier to observe than the Lagoon in light pollution. This big, 26' across glob is a bright magnitude 5.9, but the large size and its extremely loose form [Class IX (9)] make it slightly trying for smaller apertures.

It's located just 1° 18' southwest of Scorpius' brilliant Antares, so finding is assured. But go slowly if you're using a smaller instrument or if you've got a particularly bad night—this one is easy to run right over. Despite its slightly forbidding stats, I went after M4 with the 4.25-inch reflector on a good evening and was not disappointed:

> This is an impressive object in the 4-inch scope tonight, even though it's into some fairly bad sky glow. No sign of resolution, but I can make out that it's much flattened, and that has a loose, grainy appearance. It "wants" to resolve into stars.

As with M80 on the other end of the compactness scale, I went to larger aperture to see stars:

> Easily visible with direct vision in the C11. It looks very much like a dim, scattered open cluster, though, until I use averted vision. When I do "look away," its globular nature is more obvious, with a dimly glowing core area a few arc minutes in diameter. Many dim suns are resolved across its central region. All in all, I can see about 20' of cluster at 200×. There is a curious line of resolved stars that crosses the core, looking like the iris of a cat's eye, lending it its occasionally heard nickname, "The Cat's Eye Cluster."

If you get an outstandingly good night in the city and have a larger telescope at your disposal, you can also look a mere 57' northwest of M4 for a little "extra" globular, NGC 6144. This magnitude 9.1 object can be challenging in urban southern skies, but it is achievable. In most urban scopes it will appear as a small, round, dim, and utterly unresolved glow.

M6

Let's finish this jaunt with the bright, easy, and beautiful Butterfly open cluster. Magnitude 4.2 M6, the Butterfly Cluster, is 20' in extent along its major axis. It takes its name from the fact that it's composed of two roughly rectangular lobes of stars joined together.

To find M6, we travel down into the region of the Scorpion's Stinger, which is an area fairly far south in declination, meaning that if you're at higher northern latitudes you'll want to wait for culmination to view M6 if at all possible. The cluster lies to the west of a line drawn between Sagittarius' Spout Star, Gamma, and Scorpius' Stinger Stars, Lambda and Upsilon Scorpii. Move your scope into position scope midway along this imaginary line and then slew 2° to the west. You can't miss it in a 50-mm finder. This galactic cluster shines out unbelievably well from the rich background of Milky Way stars.

I liked this cluster in my C11, but this was one time when I much preferred my little Short Tube 80 refractor. A short focal length telescope can provide some wonderful wide-field views in this area, even for the urban observer.

> Relatively small and rich. Best seen in the Short Tube 80 with a 17-mm Plossl at 23×. The two butterfly-wing lobes blend into each other tonight in this scope, and it looks a little more like a rectangular patch of stars than a butterfly, but it is very nice in this instrument. I don't detect any haziness that would indicate unresolved stars, even at higher powers. Twenty stars are easily seen.

Tonight's Double Star: Antares, Alpha Scorpii

You won't need any instructions to help you find Antares. This 0.96 magnitude red supergiant is Scorpius' brightest sun, and the 16th brightest star in the sky. Antares, "Rival of Mars" and "Heart of the Scorpion," is interesting for double star fans because of its little magnitude 5.4 companion 2.6″ away. 2.6 seconds of arc is not overly tight when it comes to doubles, but the large magnitude difference between the primary star and its companion makes this a difficult observation for telescopes smaller than 6 inches. The secondary star is usually obliterated by the glare of the bright primary. Use as much aperture and magnification as you can on an evening of steady seeing, though, and you'll be rewarded by a fine sight. The strongly orange primary makes the secondary star look an outrageous shade of green, though it's actually white. I find about 11–12 inches of aperture splits Antares easily from my fairly southerly latitude of 30° north.

> It's almost time to pack it in, with the eastern sky beginning to brighten and the stars of summer diving into the west, but I can't resist slewing back to M22 for one last look. As I gaze into the deep summer night, looking long and lovingly at this cosmic relic, I keep coming back to the words of J.R.R. Tolkien, which the late Robert Burnham quoted in his Celestial Handbook as a description of this magnificent cluster: "It was as if a globe had been filled with moonlight and hung before them in a net woven of frosty stars. . . ."

Tour 3

Star Nests in Cygnus

I know when summer comes. It's obvious when I look east and see the great swan, Cygnus, winging along the horizon. This giant swan completely dominates the early/mid summer eastern sky—the other star figures in the area are dim or small. Hercules' subdued Keystone can be surprisingly hard to trace from damp and hazy city skies. The same goes for Ophiuchus; as we saw, his mostly unimpressive stars are not only hard to see in the city, their stick-figure pattern is almost shapeless. Cygnus' neighbor, Aquila the Eagle, is prominent enough, but includes little in the way of bright, deep sky wonders for urban astronomers. The rest, Lyra, Sagitta, Vulpecula, and Delphinus are tiny if identifiable and can't begin to approach the majesty of the Great Bird of the Galaxy.

What's a huge swan doing flying through the night sky, anyway? Apparently this constellation has been identified as a long-necked bird since classical times. The question is *which* swan. Some say the Bird is the pet of Queen Cassiopeia. He's also been associated with the son of the great sea-god Neptune, changed to a swan to save him from being murdered by the mighty Greek warrior Achilles. The most popular myth concerning Cygnus says that he is master musician Orpheus, torn to pieces at the hands of a band of wild maenads, female followers of the strange god Dionysus. The Olympian deities were not wholly without mercy and, it's said, placed Orpheus, wondrously changed to a magnificent bird, in heaven near his beloved harp Lyra.

But what matters to astronomers is how the real swan, the constellation stick-figure, looks and how difficult it is to find. It is remarkably easy. While this constellation's shape does somewhat resemble a winging swan if you squint your eyes and hold your mouth just right, the novice should really look for Cygnus' other guise, the Northern Cross. Cygnus, shown in the chart in Figure 7.6, forms a luminous cross stretching 22° 17′ from Deneb, the head of the cross to the beautiful double star, Albireo, at the foot. Cygnus is big, but also bright, containing stars like the remarkable Deneb, a giant A2 star burning at magnitude 1.29. In fact, all the stars that form the cross shape are easy to see from the most light-polluted urban locations, with the dimmest being magnitude 2.87 Delta, the tip of the western cross arm. The Northern Cross is arranged so that its head points north-northwest.

What's to be found in Cygnus? Hard core deep sky observers, those with semi-dark skies, anyway, tend to think of the Bird as the home of many fascinating emission nebulae. Within Cygnus' borders are the North America Nebula, the Bridal Veil Nebula, the Crescent Nebula, and many others. In bright urban skies, though, most of these tantalizing objects are invisible. They are either big and dim, or small and dim, or just plain dim. All is not lost, though. Sitting astride the Milky Way, the Cygnus region is the home of numerous open star clusters. These clusters tend not to be the best in anybody's list (though two Cygnus clusters are included in the Messier), but their sheer numbers make a visit to the Swan, a must for urban observers. And, as the telemarketers say, "That's not all!" We'll also visit a passable globular cluster lying just over the Cygnus border.

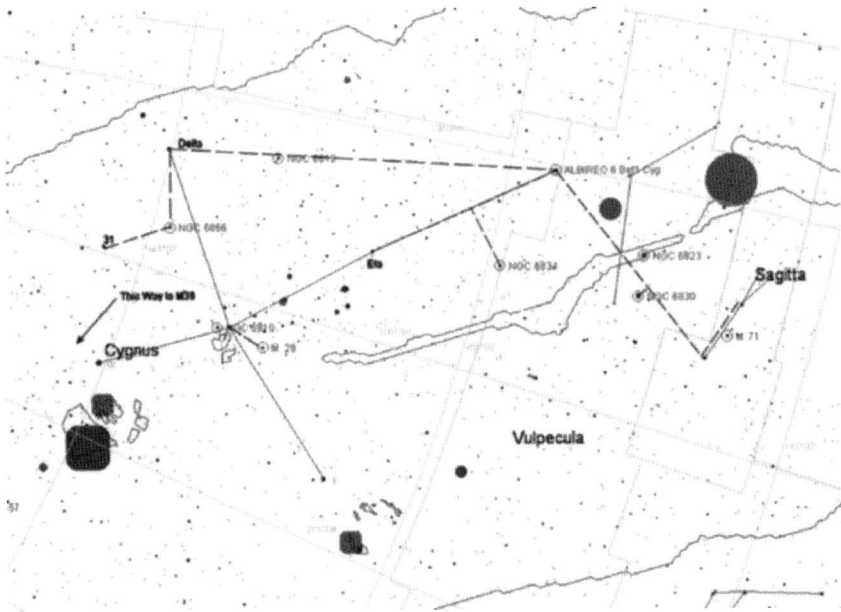

Figure 7.6. Chart of the Northern Cross.

M39

Cygnus is just *full* of open clusters and it's easy to get confused and lose your bearings. So, rather than hopping around, let's start at one end of the Swan and work our way along carefully and methodically. For our first stop, we'll begin at the head of the cross (or the tail of the swan if you prefer). M39 actually lies well away from the stick-figure pattern of Cygnus, being 9° 13′ from bright Deneb. It's in Cygnus, but just barely, lying closer to the border of the small constellation Lacerta the Lizard than to the magnificent cross. Cygnus is also adjacent to the familiar and distinctive constellation Cepheus the King, so a good way of locating M39 is by starting in that constellation rather than messing around with the dim stars of the lizard.

Our jumping-off point for M39 is magnitude 3.35 Zeta Cephei. Cepheus is not the brightest constellation in the summer sky, but his large "house" shape shows up well in fairly poor conditions. You may want to use binoculars to help orient yourself at first if he is hard to make out. Using binoculars as a navigation tool is a trick far too few deep sky observers—city or country based—know. If you're unfamiliar with an area of the sky, the bright, right-side up, wide-angle images of a pair of 7 × 50 or 10 × 50 binoculars can be a tremendous help. Once you're acquainted with the Cepheus—Cygnus border region, draw an imaginary line from Zeta Cephei to bright Deneb. Then, backtrack until you're at the halfway point of this line Zeta. Position your scope there. Then, move it 3° to the east-southeast. M39 should be readily visible as a hazy, little spot in your finder. If you need more help, the magnitude 4.23 star Pi Cygni is just over 2° 30′ northeast of the cluster.

Arriving at M39, you may or may not be disappointed depending on your perspective. It's certainly bright at an integrated magnitude of 4.6 (its brightest individual stars are at magnitude 7). Unfortunately, it's also large at 32′ and not overly rich, with 30 stars being verified as cluster members. There's no denying that what makes an open cluster interesting most of the time is small size and rich makeup. M39 is saved from total obscurity by being set in a lovely field and by having an interesting shape.

In the eyepiece, M39 is a triangle of three bright stars filled-in with quite a few dimmer cluster members. At its over half degree size, this object will challenge narrow field scopes, and it looks far better if you can put some eyepiece field around it. In other words, this is a perfect object for a short focal length rich-field refractor or reflector. Not that it doesn't look OK in longer focal length instruments:

> M39 is low in the sky and in some heavy light pollution, but it looks remarkably nice in the Nexstar 11. This large, star-spangled beauty fills the field of a 22-mm eyepiece. It gets even better in a 35-mm, though I lose some of the dimmer cluster-members to sky glow with that eyepiece. At medium magnification I can easily pick out at least 20 stars in addition to the three brightest members. These three bright luminaries are arranged in the shape of a near equilateral triangle, with the dimmer stars in this triangle's center.

But this is really a cluster for wide-field scopes:

> In the Short Tube 80 refractor at 16×, M39 comes alive and is much more interesting and attractive than it is in a narrow field instrument. The three bright "triangle" stars are readily visible, as are many of the other members. No, I can't count as many stars as I can with higher magnifications and larger scopes, but the cluster just looks much better, set in a wide and distinctive field.

Even in small-aperture binoculars, M39 is obviously composed of stars, so I assume Charles Messier, knew this was not one of his pseudo-comets when he included it in his list in 1764.

M39, as the title of this section implies is, like all open clusters, a place where stars have recently begun to shine. This particular assemblage is middle aged for an open cluster, at 270 million years. This seems ancient to us short-lived humans, but as things go in the wider universe this cluster is almost a newborn. Especially when compared to the globular clusters. Its distance of 800 light years means the triangle of M39 covers an immense 7 light years of space.

M29

The other Messier object in Cygnus is—big surprise—an open cluster. M29 is both smaller and easier to find than M39. To locate it, you start at Cygnus' "middle" star, Gamma Cygni. From there, move a mere 1° 45′ to south-southeast. This is a fairly rich region, but you should be able to identify magnitude 7.5 M29, a small fuzz-spot in your finder scope, without much difficulty. Remember, as always, to rotate Figure 7.6 (or your star atlas) so the view matches what you see in your finder.

When I look at M29, what I see is a small dipper-shaped asterism (Figure 7.7) formed by the cluster's 8 bright stars and supplemented by a number of dimmer ones. This small cluster covers about 11.0′ of eyepiece field. I used to think this cluster was uninteresting, but, like M39, it has such a distinctive shape that I keep coming back

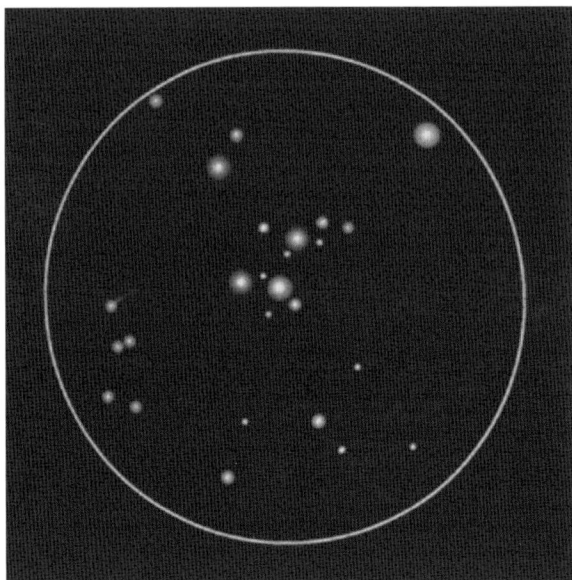

Figure 7.7. M29, open cluster in Cygnus.

to it. It has become an old friend. You can throw as much aperture at it as you want, but my observation is that it doesn't get much better after about 6 inches. Certainly, I had a good view of it even with the 4.25-inch Newtonian:

> Four bright stars stand out extremely well at 48× in the 4.25-inch scope. In addition to these four, I can see 7 other cluster members with averted vision despite scattered clouds and fairly heavy haze.

Unlike M39, this is an object for a telescope. Binoculars will reveal it as a hazy patch near Gamma, but without real resolution. When the comet-hunting Mr. Messier ran across this one in 1764, it probably did look nebulous to him in his modest telescope.

Poor M29. It has the misfortune of being set in the midst of clouds of interstellar matter that dim it considerably. Were out it in open space, this cluster would be a real standout, like a miniature Pleiades. As is, its integrated magnitude figure is at around 7.5, and its brightest stars hover at 8.5. This dimming, "extinction," means that M29, which is fairly distant from us at 7,000 light years, remains an interesting also-ran in the open cluster beauty contest.

M71

I said we weren't going to jump around. But let us do just that for a moment before resuming our survey of the Cygnus open clusters. M71 is not the area's only globular, but it is far more interesting in light-polluted skies than dim and bland M56 that lies nearby in Lyra. Situated close to Cygnus in the small constellation Sagitta the Arrow, M71 is more attractive than M56, and is both a challenge and a puzzle.

If you're patient and wait for it to rise out of the horizon murk, Sagitta becomes laughably easy to identify in most urban settings. It really does look like a small arrow, and is what I'd call "prominent" despite its brightest member being Gamma (no, not Alpha), a modest magnitude 3.47 yellowish star. Locating M71 is not at all troublesome if you can see Sagitta. M71 is positioned halfway along a line drawn between Gamma and magnitude 3.82 Delta. Actually, it's 15' Southeast of this line, but a medium-wide eyepiece should show it up as long as you *precisely* position the scope midway between Gamma and Delta.

When you think you are in the correct area, look very carefully—examine the field obsessively—and you'll notice the *small* glowing knot that is the relatively dim M71 shining at an integrated magnitude of 8.30. It is small at 6', but that won't help much. This is a *very* loose globular, so it *won't* be bright, far from it, and you'll want all the aperture you can muster for easy identification and to get a decent view of this weak and scattered cluster. I found my 11 or 12.5-inch scopes just about *required* on a poorer-than-average night. In the 12.5-inch scope it was visible if unremarkable:

> M71 looks more like a small barely resolved open cluster than a globular star cluster . I catch quite a few cluster members flicking on and off around the edges, but the core is tough. It is insanely loose and very dim. At first glance, M71 seems rather shapeless, but extended observation seems to reveal that it is triangle shaped. The field it's set in, which is beautiful and rich with stars from dark skies, is very much subdued from my light-polluted backyard.

Like most of the tougher globulars, M71 (Plate 39) is difficult because it is loose. *Incredibly* loose with a Shapley–Sawyer class of XI (11). In fact, astronomers still debate whether this group should be classified as an open or globular cluster. A check of the cluster's HR or Color Magnitude Diagram (plots of cluster stars' spectral types against their magnitudes) is ambiguous, with M71 appearing to be similar in age to older *open* clusters. It's also small for a globular. 13,000 light years of distance from us makes it only about 25 light years in diameter, a real miniature as far as these usually gigantic star balls go. These days, the consensus of opinion among professional astronomers seems to be that this *is* a globular despite appearances, though that does not appear set in stone.

More Star Nests

NGC 6910

To begin a quick tour of what's only a taste of the bright non-Messier Cygnus clusters available to urban observers, move back north to the area of Gamma Cygni where you'll find little 7' diameter magnitude 7.3 NGC 6910. Often open cluster fans run into problems when negotiating rich areas like the Cygnus star fields. The area is so crowded with stars that almost every field contains clumps of stars that *might* be the cluster being sought. NGC 6910 doesn't present that problem. In almost any scope, it's instantly identifiable and is very distinct from the background clutter.

NGC 6910 can be found peeping out only 32′ north of Gamma. There aren't any other bright stars in the area, but you shouldn't need them. Put Gamma on the Southern edge of an eyepiece yielding a half degree true field, and you'll find the cluster on or just outside the Northern edge of the eyepiece. This is a small group, and medium power will make it stand out much better than it will in the gray background of a low-power ocular. In the 8-inch Newtonian at 100×, NGC 6910 was much better than I had expected (deep sky observers have a tendency to expect NGC open clusters to be dim and/or bland, but they are not always so lackluster):

> A real surprise in the 8-inch f/5! Very nice indeed for a non-Messier. Not very rich, with about 10–15 stars visible from this site, but compact and interesting. The brightest stars form an elongated "Y" shape.

NGC 6866

Next up is NGC 6866, which is nearly the same distance down the cross as NGC 6910, but in the western section of the constellation halfway along the cross-arm toward magnitude 2.87 Delta. A magnitude of 9.0 and a major axis 6′ across mean NGC 6866 is both dimmer and smaller than NGC 6910, and it really needs a medium-sized telescope to show up well in my skies.

To find this one, use Delta and a dimmer star, magnitude 3.79 31 Cygni. 31 is 3° 19′ to the north of the cluster. Delta, on the west and a little to the south, is almost the same distance from NGC 6866 as 31 Cygni, 3° 30′. Delta and 31 form a near right triangle with the cluster. In my 12.5-inch scope, this object is lovely from city or country. The better your site, the better it looks. From country or good suburban skies, this one is particularly nice, and is set in a rich and beautiful field. If you are restricted to that good, old city observing location like many of us, don't despair, though, as you get the same effect and almost as good results by waiting for this cluster to climb to culmination:

> Beautiful field with cluster obvious and looking like a miniature M39. At about 5′ across, it is nice in the 12.5-inch scope at 200×, and presents the same effect as the Messier cluster: a triangle of prominent stars filled with dimmer sparklers.

NGC 6819

NGC 6819, another standout, is much more distinctive than the average NGC open cluster. In fact, it looks very similar to M71—a splash of stars with a slightly compressed, hazy core. This object can be a little tough to find, since it doesn't live near any prominent stars. The easiest way to locate it, probably, is to position the scope so that it points to a spot two-third of the way along a line from Beta Cygni, Albireo, to Delta. You'll find it 5° south and east of bright Delta. Another way of looking at the situation is that the cluster forms a tall triangle with Gamma and Delta. The brightest star near NGC 6819 is magnitude 4.89 15 Cygni, which lies back along the line to Albireo, 3° from our target cluster. Search carefully, since it's only magnitude 9.5. Its small 5′ size does help here, as it may even be visible as a little haze spot in a 50-mm or larger finder.

> A very attractive NGC open cluster in the 11-inch Schmidt-Cassegrain. I started out at 127×, which revealed the cluster as a squarish pattern of very tiny stars about 5′ across. At higher power, the cluster grew a bit, as more dim stars were revealed, and it looked more oval than square. Tonight, this one actually looked better than M71.

NGC 6834

This open cluster is in the southern part of Cygnus near Albireo in the foot of the cross area. NGC 6834 is east of a line drawn from magnitude 3.89 Eta to brilliant magnitude 3.08 Beta. Move the scope to a point midway along this line, and then slew 2° 30′ east and slightly south and you should be able to get this somewhat dim 9.7 magnitude 5′ diameter cluster into the field of a medium-powered ocular. If NGC 6866 looked better in medium apertures, this cluster cries out for it, with my best view coming with the Nexstar 11.

> Small and dim. In the 11-inch scope I see a 5′ oval of faint stars—mostly magnitude 10 and dimmer—crossed by a prominent line of brighter stars. Until I increased the magnification to 220×, I had the impression that nebulosity was involved in the cluster. At higher power this is revealed to be many faint stars.

NGC 6830

The neighboring constellation Vulpecula's NGC 6830, at magnitude 8.0 and 12′ in extent, shouldn't be easy to separate from the background star fields. In fact, many observers report that it is invisible against the rich Cygnus Milky Way background. I tend to think that this is really a case of insufficient aperture, however, as this one was impressive and easy to see in 11- and 12.5-inch instruments. If you have access to at least a 10-inch scope, this NGC is certainly worth searching for.

NGC 6830 lurks in Eastern Vulpecula near the border with Sagitta. The best pointer is magnitude 4.58 13 Vulpeculae, which is 1° 7′ north and slightly east of the cluster. While 19 will be visible in a 50-mm finder, you will likely not see it naked eye from the city. If you have trouble positioning the scope correctly, another way to find NGC 6830 is to move the telescope so that it is about 2/3 of the way along a line between Beta Cygni and the "Arrowhead," Gamma Sagittae. For best results here, choose an eyepiece that will encompass the cluster's 12′ size and allow some space around it, while delivering medium-high magnification. I did this and had no trouble finding or observing the cluster:

> NGC 6830 is another good one for medium aperture. Very distinct from the rich, beautiful field it is set in. Rectangular in shape with three brighter stars and many dimmer ones, i.e., 10–12 dimmer stars easy.

NGC 6823

While I've suggested medium aperture for the two preceding clusters, they are no doubt visible—for a persistent observer—with an 8-inch or smaller telescope. They may not be easy, however. That's not the case with NGC 6823, "Scorpius Junior," as I've christened it. It's fairly large at 12′, but its integrated magnitude, 7, is much more "reasonable."

If you were able to locate NGC 6830, this one will be a breeze, as it's only 1° 48′ west of the preceding open cluster. If you have trouble, you can also pin it down by aiming the scope so it's almost exactly halfway along a line that runs from Beta Cygni to magnitude 3.82 Delta Sagittae.

> A nice, medium-sized open cluster in the 8-inch f/5. Looks very much like a miniature Scorpius. Without much imagination, I can visualize the stinger, tail, and head of Scorpius' little brother. Six bright stars stand out and numerous dimmer ones wink in and out of view.

Some observers have reported glimpsing Pleiades-like nebulosity involved with this cluster, but I saw no trace of it at any magnification with or without LPR filters.

Tonight's Double Star: Albireo

Yes, this has been the *barest* sampling of the Cygnus area's open clusters. There are *many* more interesting ones that deserve a visit. But I've had enough for the night. Heavy midsummer dew is falling and a warm bed is calling. I'll make just one more stop before tearing down the scope: Albireo. This wondrous magnitude 3.1 double star, Jewel of the Swan, is a not to be missed sight. Composed of two stars 33″ apart, the primary is a deep golden yellow while the secondary is an incredible azure-blue. Novice or old hand, *nobody* can possibly grow tired of Albireo. This is one time when aperture is not a consideration. I thought Beta looked as good (or better) in my trusty 80-mm *f*/5 refractor as it did in the 12.5-inch Dobsonian.

> The incredible beauty, not just of gem-like Albireo, but of all I've seen on this night and a lifetime of other starlit nights, literally tugs at my heartstrings. How incredibly lucky I am to be an amateur astronomer. To think, if I hadn't stumbled across that dog-eared Patrick Moore book in my elementary school library I'd have missed all this. Maybe that's why I'm so interested in introducing new people to the sky. We amateurs are, by action and example, purveyors of wonder, evangelists of the cosmos—we just can't help ourselves.

Tour 4
Requiem for the Dead Stars

Stars are like people. They are born, live their lives, and die. The courses of their lives are more majestic outwardly than those of humble creatures like us, of course. They are birthed in light-years-spanning clouds of gas, live in thermonuclear fire, and some of them die in supernova glory that can outshine an entire galaxy for a brief time. But like most of us, the majority of stars live sedate lives. The fate of the smaller stars, stars like our own Sun, is less majestic than the fate of the giants who die as immense fireballs. The little suns die with a whimper rather than a bang, but leave behind remains as fascinating as a supernova and its remnant nebula.

The corpses of Sun-class stars are far more common and easier to see than the remnant nebulae of supernovae. These supernova remnants like the diaphanous Bridal Veil Nebula in Cygnus can be dim and difficult in small scopes from dark country and are nearly invisible in the midst of city light pollution.

The little stars, whose lives end as red giants, slowly blow off their outer layers and leave behind planetary nebulae. Planetary nebulae have nothing to do with planets. They are called "planetaries" because Sir William Herschel, who ran across many of these objects in his restless sweeps, his surveys of the night sky, thought their fuzzy disks somewhat resembled "his" planet, distant Uranus.

When a smaller star ends its life, it expands to red-giant size as nuclear fusion in its core shuts down as hydrogen is exhausted. Over the course of millions of years, the huge clouds of gas that once formed the star's atmosphere slowly drift off into space, revealing its slowly cooling core, the leftover body of a once mighty sun ending life as a dimming white dwarf. This core still puts out a great deal of radiation all across the spectrum, however, and excites the drifting clouds of gas to glow brightly. Voila! You have a planetary nebula. These are prime objects for the urban astronomer, since most are small and bright compared to either diffuse nebulae or supernova remnants. There are challenging planetaries, but the wonderful words "small" and "bright" do indeed describe the appearances of the majority of these objects.

Planetary nebulae are especially welcome targets during the summer. Open star clusters are nice, but observing one after another all summer long gets boring—for me anyway. In contrast, planetaries are almost always surprising. The events that take place at the end of a star's life usually leave the white dwarf sitting in something that resembles a tube of gas. The appearance of this gas tube depends on our perspective. If we're viewing it end-on, it will look like a torus, a donut. If it's arranged so we see its side, it may look like a rectangle or box. Depending on a number of factors, planetaries can assume some almost fantastic shapes. The Dumbbell Nebula in Vulpecula actually resembles a half-eaten apple, while Scorpius' Bug Nebula looks like an insect.

Where do you go to find planetary nebulae? They are dead stars, so you go where the stars congregate in their greatest numbers, along the Milky Way. The entire stretch of the summer Milky Way from Cassiopeia in the North to Sagittarius in the South is littered with the bodies of the dead. There are so many planetaries gracing the summer sky that even a brief survey of what's out there would take many, many a summer night, even from urban sites. For our guided tour I've selected those that are both bright and not too small or large.

M57

This is as good as it gets. The famous Ring Nebula, M57, is 1.4' × 1' across its major and minor axes, just the right size to stand out well from the bright city sky background. Due to its small size, it is amazingly bright at magnitude 9.4. While this object gets more and more impressive with increasing telescope aperture, it is a fine sight in an 80-mm refractor and easy enough in a 60-mm. To find it, you'll visit the small and distinctive constellation Lyra, a classical star pattern that represents Orpheus' heaven-preserved harp. Lyra would be fairly nondescript were it not for its Alpha star, magnitude .03 Vega, a member, with Deneb and Altair, of the famous "Summer Triangle" of bright stars. Vega is a mighty class A0 star, a blue–white beauty that's one of the gems of the northern sky.

M57 is the easiest of tonight's objects to locate. It's on the opposite side of the lyre stick-figure from Vega, and is found a little less than halfway along a line drawn from Beta Lyrae to Gamma Lyrae. Position the scope 3/4 of a degree, 48' to be exact, from Beta directly on the line from Beta to Gamma. Magnification doesn't matter too much, but you probably want to be at around 50× or thereabouts.

I found that if I went too low—16× in my Short Tube 80 refractor, for example—the Ring stopped showing up as a ring and looked more like a dim and slightly fuzzy star. Aperture seems important for the best view of the Ring, too. Increasing the power did make M57 distinguishable from a star in the little refractor, but even at higher magnifications it was hard to see the nebula's smoke-ring shape with the 80-mm or with 4-inch class telescopes. I did find that averted vision almost always showed the "donut hole" with my Short Tube 80 at 100×, but it wasn't an easy observation with that scope or with the 4.25-inch Newtonian:

> I remembered M57 as being slightly difficult in heavy light pollution in this small New-tonian, but it's very prominent tonight in far less than perfect skies. Initially appeared as a perfectly round fuzzball, but averted vision and a magnification of 90× showed the ring shape with some difficulty—enough difficulty that I don't think a novice observer would have seen it. Much of the time it looked like an undistinguished, dim gray disk.

At 6 inches of aperture, the nebula's ring aspect becomes easy to see with direct vision. Aside from greater brightness, M57's appearance doesn't change much more with increasing telescope aperture until you get to 10–12 inches. A scope of that size in the city begins to reveal more of the Ring's secrets, like the fact that it isn't a perfectly round donut, but is elongated. It also becomes clear that this is really a "filled" donut. The interior hole is not dark; it is an obvious light gray. What's the holy grail of Ring observers? The central star. The white dwarfs that form the central stars of planetaries are often easily visible in small scopes—but not this one. The Ring's dwarf is exceedingly dim, around magnitude 14 or *dimmer*, and the hazy nature of the ring interior makes it even more difficult to see.

I have never seen M57's central star from the city in any scope, including a 24-inch monster Dobsonian, and have seen it only with *extreme* difficulty in a 12.5-inch scope at very high magnification under very dark and steady skies. That doesn't mean you shouldn't attempt it, however. Use the largest scope possible, and the highest possible magnification—500× and above—to thin out the interior nebulosity and create enough contrast for the star to emerge. You'll also need rock-steady seeing, as unsettled air that makes stars "bloat" and shimmer will completely erase the Ring's central star.

If you thought distances to other DSOs sounded uncertain, you haven't seen anything yet. Like many—if not most—planetaries, the distance determinations for M57 are all over the map, ranging from 1500 to over 5000 light years. Even its shape is a matter of speculation. It used to be thought that the Ring was a sphere, with thicker gas at the limb causing the ring appearance. Today, it's believed that M57 actually *is* a ring, or, perhaps more likely, a tube of nebulosity viewed end-on.

M27

There's no denying the Ring is a real classic of a DSO, but its relatively small size and fairly regular ring shape argue against it being considered the best planetary in the summer sky. There is not a whole lot of detail for the small-medium scope to pick out in M57. Many observers will tell you that the "best" honor should go to another summer treat, M27, the Dumbbell Nebula, located in the unassuming constellation of Vulpecula, The Little Fox, which we visited earlier in search of an open cluster. Magnitude 7.3 M27 is bright and also big at $8' \times 5.7'$, making it an interesting target for giant binoculars. While it's large in comparison to the Ring, it's not large enough that its light is so badly spread out as to make it a challenge from the city.

Vulpecula may not be a familiar constellation for novice astronomers, and it's certainly not prominent. Its three brightest stars form a distorted triangle (Figure 7.8) that's hardly eye-catching. The brightest star in the constellation, Alpha Vulpeculae, is a dismal magnitude 4.4. The Dumbbell should be hard to find, then? Not at all. The proximity of Vulpecula to glorious Cygnus and distinctive little Sagitta means it's easy to get your bearings and track down this DSO. Once you're in the general vicinity, you'll find plenty of guide stars, too—here near the Milky Way, many stars show up with only a little optical aid, even in bright skies.

There are several approaches that will land you on M27, but the route I usually take is from magnitude 3.47 Gamma Sagittae, the "Arrowhead". Draw an imaginary line from Gamma to magnitude 4.58 13 Vulpeculae 4° 45' away. 13 should be easy in most finderscopes. M27 lies 3° 13' along this line toward 13 and just a bit outside the line to the Northeast. This is an area lacking in bright stars, but magnitude 5.67 14 Vulpeculae, just 23' Northwest of the nebula, is prominent in finders, and is an almost infallible guide to the Dumbbell.

M27's nickname the, "Dumbbell," goes all the way back to William Herschel's son, John, who likened its double lobed appearance to a barbell. This double-lobe shape will be obvious in the city if you're observing with at least a 6–8-inch telescope. If you're using a 4-inch or smaller instrument, the dumbbell shape is not always easy to detect. You'll probably see something more like the strongly oval fuzzy in Figure 7.9, a drawing I made some years ago from a highly polluted urban site with the 4.25-inch

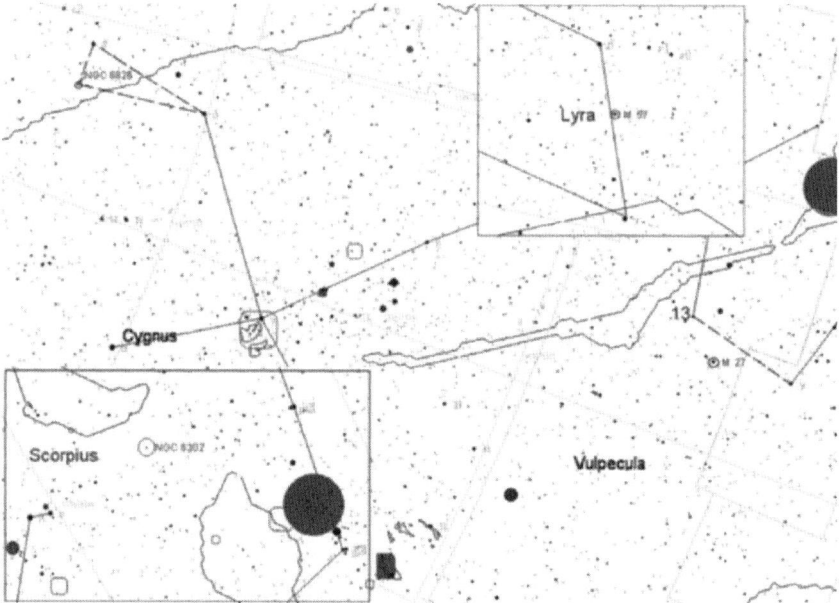

Figure 7.8. Vulpecula and companion constellations.

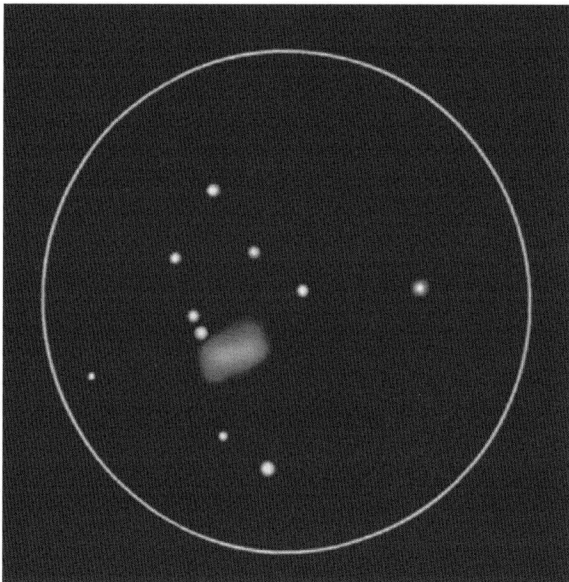

Figure 7.9. The Dumbbell Nebula as seen in a 4-inch scope.

Newtonian. On that evening, the Dumbbell looked a lot like the Crab Nebula does from a dark site with a 6-inch telescope.

In my little 4-inch scope, M27 was:

> Fairly easy to identify, but the dumbbell shape is extremely elusive, being barely perceptible at times. I have to work to see anything other than a round-appearing nebula. Averted vision helps, but higher magnification does not seem to. M27 is a dim gray in color with diffuse edges. This is an easy object in the city with an 8-inch telescope, but I missed it a couple of times with the 4.25 before finding it.

If you're the owner of a UHC or OIII filter, by all means use the filter. The OIII, in particular, will help M27 tremendously. It will make the nebula's apple core shape easier in a small scope and unmistakable in a mid-sized instrument. Paradoxically, when you increase your aperture to about 12 inches and insert an OIII, M27 begins to lose some of its famous shape. More outlying nebulosity comes into view and the Dumbbell begins to morph into another piece of sports gear, an American football. Given M27's medium size, use medium powers. 100× is a good level to begin at when you're hunting details.

And there are many details to be had, at least in 8-inch and larger apertures. One of the easiest to detect is a diagonal bar of brighter nebulosity that runs from lobe to lobe. Careful study of the nebula at higher powers will reveal a number of other bright patches in the central region, too. You'll also notice if you remove the OIII filter—numerous dim stars scattered across M27's face. If you leave the filter off, you may even have a chance of locating the central star.

Unlike the Ring's dauntingly dim precursor, the central star of the Dumbbell is quite approachable at magnitude 13.5. The secrets to spotting it are using high power to dim down the surrounding nebulosity and using a photograph or chart with the central star labeled to help you pick it out from the many faint stars scattered across the nebula. M27, at a fairly sure distance of 1200 light years, rates as "close" when it comes to planetaries, and that is responsible for the wonderful view we have of this star-corpse from town.

M27 is one of those objects, like Orion's M42, that I never tire of, and come back to summer after summer. I've observed it hundreds of times over the course of my observing career, and imaged it almost as often. It shows up well in CCD pictures taken from the worst light pollution, and I captured it recently (Plate 40) with a Starlight Xpress MX516 camera and my C8 SCT on a night when it was almost "not there" in the eyepiece. This fabulous object may present us with a picture of what the Ring Nebula would look like if viewed side-on rather than from the end.

NGC 6826, The Blinking Planetary

Almost as attractive and interesting as the Dumbbell is a slightly less well-known summer object, the Blinking Planetary, NGC 6826. This is a planetary nebula, like M27, that will reward all scope users in the city or in the country. It is dimmer than

M27 at a magnitude of 8.8, but it's also, like M57, small at 24″, making it very bright. I've had no trouble distinguishing it from a star with a 60-mm ETX at relatively high magnification despite the nebula's small size. Large scopes with OIII filters and high powers begin to reveal some interior detail in the nebula, but what most people remember is the amazing "blinking" effect.

The thing that sets NGC 6826 apart from both the Ring and the Dumbbell is the brilliance of its magnitude 10.6 central star. It's impossible to see the Ring's central star with a small scope, and something of a struggle to pin down M27's dwarf, but the Blinking Planetary's remnant star is often the most obvious feature of this object in a small telescope. The nebulosity is bright, too, but a little harder to see than the star. In fact, for the nebulosity to show up well in 8-inch and smaller scopes in the city, the observer will have to use averted vision. Stare straight at this object, and all you see is the star. Look away and the nebulosity pops into view. Rapidly switch from looking directly at NGC 6826 to looking off to the side *and the nebulosity blinks on and off.*

I think the Blinking planetary is definitely easier to find than the Dumbbell. Being located in Cygnus, there are, as was the case with M27, plenty of guide stars. But it's also well-placed with respect to prominent sparklers, not just nondescript 4th and 5th magnitude suns. My method for locating NGC 6826 is a simple one. The blinking planetary forms an elongated triangle with the magnitude 2.87 Delta Cygni, the tip of the western arm of the cross, and magnitude 4.48 Theta. The nebula forms the base of the triangle with Theta, and is 1° 22′ east of the star. A potential difficulty may be in locating Theta, but it's fairly prominent in a finder, and is exactly 5° 17′ Northwest of Delta Cygni.

When the Blinking Planetary is in the center of your field, increase magnification to 150× or higher for a nice image scale and play around with the "blinking" effect for a while. You may be surprised to find the nebulosity blinks more readily with small scopes than with large ones. Larger than 8-inch instruments begin to pull out more and more nebulosity; so much nebulosity that at least some of it is visible with direct vision, lessening the dramatic blinking effect. This nebula responds well to an OIII filter, but using one will *eliminate* the blinking. The OIII both suppresses the central star and brings the nebulosity into direct-vision range of even smaller telescopes.

I did mention details earlier, and some *are* there in this object for medium and large scope owners. In a 12-inch or larger aperture, two bright nebulous patches, one on either side of the central star, may be seen. At high powers in OIII filtered scopes, an inner ring structure may also be discernable.

What can the small scope owner expect? My log from a hot August night using the 6-inch Newtonian is a pretty accurate depiction of what's in store with 4–8-inch telescopes in the city. The nebulosity will be slightly dimmer in the 4-inch scope and brighter in the 8-inch scope, but I didn't begin to see anything other than the pretty blinking effect—any details—until I got to 12 inches.

> Round with a fairly distinct blue color. The blinking is more prominent at magnifications of 100× and higher, which helps pull the central star out of the light pollution. The disk of nebulosity around the star appears to be smooth and even in this telescope.

More Dear Departed

NGC 6302

Cygnus is not the only area of the Summer Milky Way peppered with planetary nebulae. Aquila and Ophiuchus possess many of these nebulae, though no really outstanding examples. Get down into Sagittarius and Scorpius and things improve dramatically. For my money, one of the strangest looking and most interesting planetary nebula in the sky is Scorpius' odd little Bug, NGC 6302 (Plate 41). It is located almost halfway along a line between one of the Stinger Stars, magnitude 1.58 Lambda Scorpii and magnitude 3 Mu Scorpii. While some references list this as a diffuse nebula, its planetary character is pretty clear to professional astronomers.

The Bug is dim at magnitude 12.8, but its size of 1.2′ × 0.5′ means high surface brightness. What's most remarkable is that even smaller scopes show this planetary's odd shape. It's composed of two lobes, not unlike a miniature of the Dumbbell Nebula (it's sometimes referred to by the alternate name of "The Bipolar Nebula"), but that's not the overall impression I have of this object, not in my 11-inch and larger telescopes. The bi-lobed structure, combined with barely detectable filaments and a flaring on one end, make this look exactly like a little ant crawling across the eyepiece field. "Bug" indeed!

NGC 6543

Not all planetary nebulae are located in the Southern Milky Way. Stars are everywhere in the sky, so, their remains can also be found anywhere in the sky. Our final stop for this evening, the magnitude 8.3 Cat's Eye Nebula, NGC 6543, is famous due to an incredible Hubble Space Telescope image. In most amateur scopes, however, it looks far blander, more like the image in Plate 42. This object was, by the way, well known among deep sky observers long before the HST was pointed its way. Located in Draco, it's at its best on August nights when it's nearing culmination (assuming you're luckier than I am and actually experience clear weather in August), and is easy to locate near a magnitude 3.17 star, Zeta Draconis and a magnitude 3.0 sun, Delta Draconis. Draw a line from Zeta to Delta. The Cat's Eye's small 24″ disk will be found 5° 6′ from Zeta in the direction of Delta, and just a hair to the south of the line.

Large amateur scopes, high magnifications, and dark skies can reveal some internal detail in this object. Certainly not anything close to what's in the Hubble shot, but at least some indications of bright patches or streamers near the central star. Me? Beyond a vaguely oval cat-cye shape, an obvious central star (magnitude 9.5), and a noticeable bluish-green coloration, I rarely see anything that could be called "detail" in this nebula in any of my scopes, even from the darkest sites. This is a bright, interesting, and magnificent object at magnitude 11.27 (which, shouldn't deter you— it's small, remember) despite a slightly bland appearance, and deserves a look every summer evening. If you're on a quest for details, good luck. Your best bet is to use large aperture and high magnifications—500× and above—on steady nights. The weird and wonderful filaments and helical structures visible in the HST images are now thought to be the result of the progenitor star having a companion, that is, that the central star is actually a double star system.

Tonight's Double Star: The Double Double, Epsilon Lyrae

Back over in the distinctive little constellation of Lyra, on the opposite end of its stick figure from M57, you'll find a truly marvelous double star system, Epsilon Lyrae, the famous Double Double. It's called the "Double Double" because it is made up of two *pairs* of double stars, Epsilon[1] and Epsilon[2] separated by 208″. Epsilon[1] is composed of a magnitude 5.0 primary and a magnitude 6.0 secondary separated by 2.6″. Epsilon[2], which lies to the south of Epsilon[1], is a magnitude 5.2 primary and a magnitude 5.5 secondary separated by 2.3″. All four stars are white in color.

As you'd expect, it's easy to separate Epsilon[1] and Epsilon[2] in any telescope. Resolving the components of each pair is not so easy, however. You're helped by the fact that there are no huge magnitude differences, but the 2.3″ of the second pair is clearly going to be a challenge for small scopes. Six inches can split all four stars at high power, but good seeing helps, and a larger scope is a good option if you're interested in actually seeing all the components on any given evening.

> Late on an August evening, standing out under open sky with my telescope, I sense the season beginning to die. Summer warmth left my bones hours ago, and the dew and an early morning breeze start me shivering. It's always best to dress warmly and in layers when you're out observing, but on those nights when I visit the stellar graveyards that extend all down the Milky Way, horizon to horizon, even a warm sweater doesn't repress all the shivers.

CHAPTER EIGHT

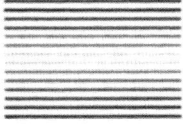

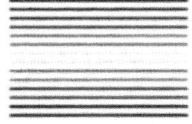

Autumn

Tour 1

A Trio of Fall Globulars

As warm September days pass and cool October nights arrive, the heavens change again in their never-ending cycle. The beauties of summer—Cygnus, Hercules, Sagittarius—are still on view but are descending into the west. If you want to catch the Summer Milky way, now's the time. It will be gone all too soon. Of course, there are many fascinating objects on the rise in the east, too. That's a big attraction for observers at this time of year. You get the best of both worlds. Point your scope west and you can still visit the multitudinous clusters and nebulae of the Milky Way. Head east and, just as in the spring, you'll be looking out of our home spiral into the galactic wilderness beyond. Many of the galaxies of fall, like Andromeda's M31 and company, Pegasus' NGC 7331, and Cetus' brilliant M77, are easily seen by city scopes. Much more spectacular than these galaxies, however, are autumn's fantastic globular star clusters.

Yes, globular clusters. Most observers think of these as summer objects, but their locations out in the galactic halo mean that just as a few—M3 and M53 among the Messiers—dip into the spring, some are also on display in fall skies. One cliché I keep hearing is "all globular clusters look alike, see one and you've seen them all." When I hear this, I can't help but think that the person making this claim hasn't observed many globulars. To the glob-fan, these star clusters are as distinct and individual as human friends, as we'll see tonight.

On this evening's star-hike, we'll visit three memorable globular clusters, M15, M2, and M56, all dramatically different objects, ranging from "blazing" to "subdued." Before we begin, though, I see that Hercules is getting low in the west. I'm sure you'll want to take one last look at his amazing star-ball. Go ahead, point the telescope at M13. I'll wait.

M15

Finally had enough of M13? Let's begin our journey, then. Our first stop is M15 in Pegasus. This is a most attractive object for small scopes, including those sited under badly light-polluted skies. Dark skies and large apertures transition this one from "attractive" to "spectacular," but I've had nice looks at M15 with scopes as small as my 60-mm ETX in the city.

You won't have to spend much time hunting M15. Like all tonight's objects, it's amazingly easy to find. It is bright and prominent with a magnitude of 6.4 and a diameter of 12.3′, and lies only 4° 10′ from prominent Enif, Epsilon Pegasi, The Horse's Nose. To hit M15, as shown in Figure 8.1 draw an imaginary line from Theta Pegasi, which is the star just east of Enif in the Horse's Neck, through Enif, and on for approximately 4°. Position your telescope in this spot and look for a fairly distinct 6th

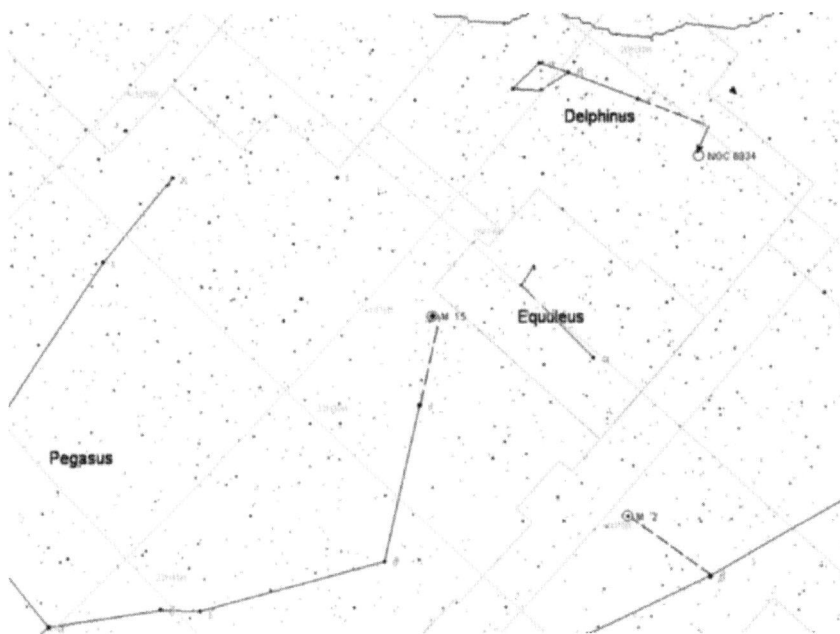

Figure 8.1. Forelegs of Pegasus, the Flying Horse.

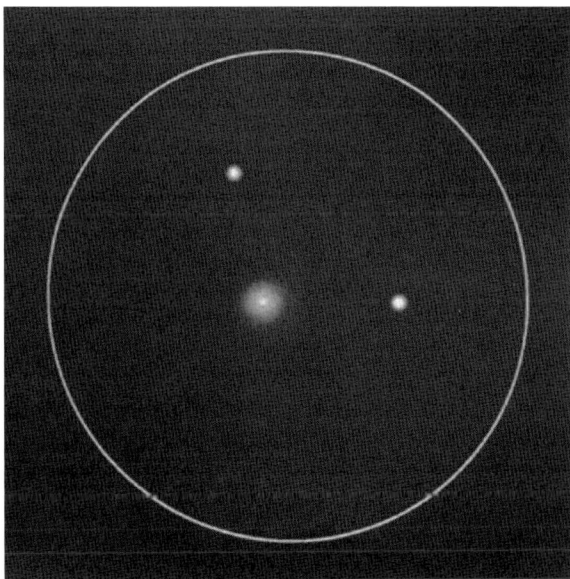

Figure 8.2. Pegasus' compact globular star cluster M15.

magnitude star. M15 is just 16′ to the west-southwest. Center this star in the finder, insert a low-power eyepiece, and take a look. With maybe a little sweeping, M15 should jump into your field.

In the main eyepiece of nearly any telescope you should notice an obviously fuzzy "star" right away. This is M15, which possesses a core that is strangely bright and compact, making the cluster show up well in the brightest skies. Increase your magnification to 100–150×, and settle in for a good, long look. Unfortunately, under the poorest skies with telescopes 6 inches in aperture and smaller, all you may see *is* M15's preternaturally bright core. There will probably also be a little haze surrounding this core, but don't expect to easily resolve stars. Careful application of averted vision and high magnification *may* make a few wink into view now and then, but you will probably need at least 8 inches of aperture to see this as a resolved globular cluster. An 8-inch scope does do that handily:

> In the 8-inch f/5 Newtonian, M15 is beautifully resolved at 166× with a Chinese 6-mm 60° apparent field eyepiece. The cluster stars, which are incredibly tiny and delicate, extend at least 75% of the way out to the eyepiece field edge, quite an accomplishment for this inexpensive telescope, as this short focal length eyepiece features over one-third degree of true field.

Even if your telescope is so small that you're unable to see many or any of this highly compressed cluster's stars, M15 is still worthy of extended observation or even a drawing, as the one I did of this glob in Figure 8.2 shows. I also made a text entry in my log on the same evening:

M15 is bright and easily found with the 4.25" Newtonian. Far brighter in this aperture than you'd expect. Seems more or less round with perhaps a hint of elongation north/south. The concentrated core is almost star-like, even at higher magnifications. Some hints of mottling, as if the cluster "wants" to resolve, but won't quite do it. I detect, at most, a star or two at the cluster edges, and can only see these with extended observation using averted vision at 200×, which is the limit for this telescope's optics.

M15 was a marvel in the C11 SCT:

Incredibly beautiful with a 25-mm Plossl in the C11, even though it's not completely dark yet. The core is bright, blazingly bright, and hordes of tiny stars are all around for about 10 arc minutes. It is just about as beautiful in a 12-mm Nagler at 220×. Resolution close to the core is very evident in this eyepiece.

M15 is a Shapley–Sawyer Class IV (4) globular, which makes it "highly concentrated" according to that classification scheme. One thing's sure; its core is amazingly luminous. Why? Current thinking is that a black hole may reside at M15's heart. Professional astronomers have changed their minds on this at least once before, however. Like so many of the objects we observe, M15 remains a mystery, its bizarrely bright center shining across the thousands of dark light years and into our tiny scopes.

M2

Once you're ready to move on from M15—and I hope you give it at least a half hour of your time—we'll set the course of our imaginary starship for destination two, M2. M2 is an impressive globular star cluster that has the misfortune of being located inside the borders of the dim zodiacal constellation Aquarius, The Water Bearer. Due to its position along the zodiac, Aquarius is a familiar name to even the most novice astronomer, but that doesn't mean they've actually seen it. Its pattern, composed of mostly lackluster stars, can definitely be hard to make out in the city.

Is M2 hard to find, then? Not very. M2 itself is bright (magnitude 6.5, 8.0′ in size), making it dramatically apparent in my little ETX refractor in heavy light pollution. It is also conveniently located near one of Aquarius' few respectably bright stars, magnitude 2.95 Beta Aquarii, which is shown in Figure 8.3. Center your finder on Beta, and then move 4° 46′ north, as shown on the chart. A pair of magnitude 6 stars half a degree apart lies a degree north-northeast of M2, and provides a good guide if you get lost in this star-poor area. M2 *may* be visible in a 50-mm finder, depending on your conditions. Even if you can't make out M2 in the finder, with a little luck it should be in the field when you move to the main eyepiece if you've positioned the scope with care. As always, have a good star atlas on hand to supplement the charts in this book.

When I had M2 firmly centered in the 4.25-inch Newtonian, I found it to be "tantalizing." I couldn't make out any individual stars, not with averted vision, and not with high power. But it was definitely granular, and seemed *ready* to resolve. This impression was much stronger than with M15:

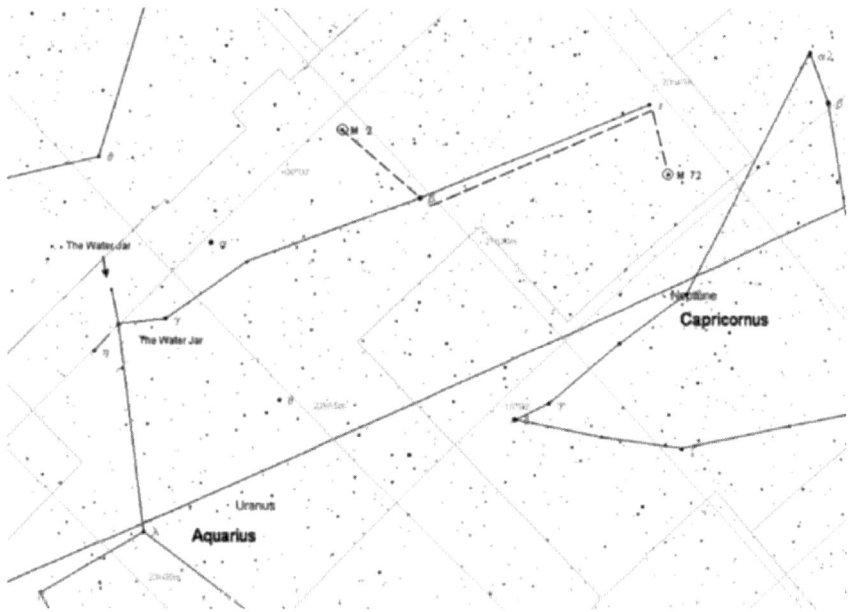

Figure 8.3. Aquarius, the Water Bearer.

> Spectacular is the word! Maybe not a beautiful as M15, but lovely nevertheless. Much less concentrated than M15 but still fairly condensed looking. Loose, grainy, and ready to resolve on a better night.

Well, maybe. I don't believe I've ever had an urban evening that was good enough to allow M2 to be resolved in a 4-inch class telescope. A little more aperture works wonders on this glob, though.

A 6-inch scope *begins* to reveal M2's stars in the city, which is not surprising, since the brightest are at magnitude 13. An 8-inch telescope delivers many more, easily and routinely. Nothing, of course, does more good than dark skies, and I had a mind-blowing view of this cluster one October evening from the dark northern Mississippi skies of the Mid-South Star Gaze. In my 8-inch Schmidt Cassegrain, M2 was revealed as a titanic globe of stars, and took on an almost three-dimensional appearance. At times, I felt as if I were in danger of *falling into the cluster*. M2 is rather strongly elliptical in shape, but it always looks round to me, no matter how large my scope or how dark the skies.

M2's core appears much more normal than that of M15, so it's a surprise to find that this globular actually rated as a *more* condensed II (2) on the Shapley–Sawyer scale. Despite its "looser" class of IV, M15 *looks* more compact because of the unusual brightness of its nuclear region. A presumed distance of 40,000 light years gives M2 a diameter of 175 light years. Estimates of its star-population density range upward from 150,000 stars.

M56

The last featured attraction for tonight is one of the more obscure Messier objects. It's really a "summer" object, I suppose, but it's sedate enough that, hypnotized by the gleaming deep sky marvels of summertime, I usually don't get around to M56 until it's almost too late. Everybody knows and loves Lyra's M57, the Ring Nebula, but relatively few people bother to visit its neighbor, M56. The lack of attention given this globular cluster is due to its fairly dim magnitude, 8.3, its rather loose structure, and the fact that it is relatively distant at 50,000 light years. It's of medium size for a globular, 7', but this comparatively small diameter doesn't seem to help with its visibility in light pollution. It can be hard in small apertures, and doesn't begin to be very interesting until you apply 10–12 inches of telescope mirror to it.

M56 is M57's neighbor both on the Messier list and in the sky. It is easy to locate by drawing an imaginary line from Beta Cygni, the famous double star, Albireo, and bright Gamma Lyrae. You'll find this globular 3° 48' from Albireo right on the line to Gamma. Work slowly and methodically, because this one will be easy to miss in light-polluted skies. *Very* easy to miss. If you have trouble, there's a magnitude 5.85 star just 25' past M56 in the direction of Gamma Lyrae.

I found M56 to be a surprising challenge when I was doing the Messier list with my 4.25-inch scope in the city. In fact, it was one of those objects that I had to search for over the course of quite a few evenings, waiting for an especially good night. I was prepared to undertake this kind of a hunt for the dimmer Messier galaxies, but was surprised and a little put out to find a supposedly bright glob so vexing. When it finally appeared in the 4.25-inch scope, I was a little less than bowled-over. Nevertheless, since I'd hunted for this one for so long, I did document it with the drawing in Figure 8.4 and a log entry, just to prove I'd been there:

> Amorphous and quite a bit dimmer than I expected. At 90×, it's nothing more than a vague and undefined glow in the middle of my field. Not even a hint of resolution. Even riding high in the sky, this cluster is undetectable until I use averted vision.

As with M2 and M15, M56 was dramatically improved in larger aperture scopes. In the 12.5-inch Dobsonian, it was much easier to find, and, when found, yielded quite a few stars, and was beginning to look a lot more like the photograph in Plate 43. Moving that scope out to our hardly perfect suburbs actually made M56 into a beautiful object, with many tiny stars resolved across its face. Unfortunately, a dim and loose structure prevents this cluster from being a real showpiece, even from dark desert locations.

M56, a Shapley–Sawyer Class X (10) globular, was one of those objects actually discovered by Charles Messier who first laid eyes on it in 1779. At an estimated distance of 33,000 light years, it stretches across about 85 light years of space. In addition to its loose structure, most of its stars are at around magnitude 14, making the cluster problematical for smaller amateur telescopes when it comes to resolution.

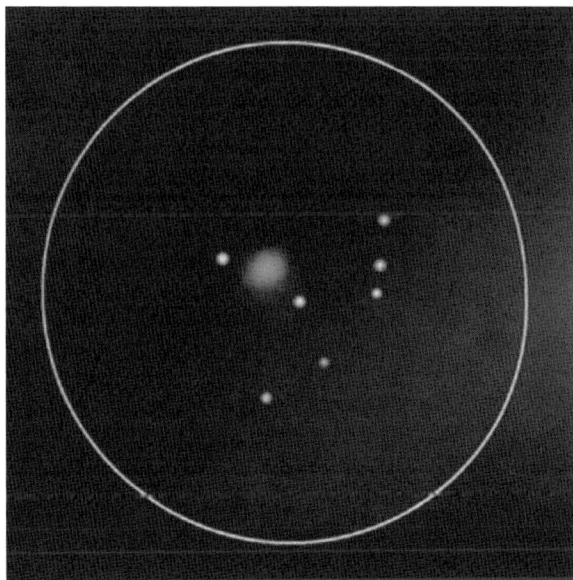

Figure 8.4. Lyra's subdued globular, M56.

A Couple More

NGC 6934

If you hurry, you can still catch the brighter of Delphinus the Dolphin's two globulars, NGC 6934. In this case, "brighter" is a relative term. At magnitude 8.9 and 5.9′ across, this little fuzzy can be difficult. It's easily located by drawing a line from magnitude 3.77 Alpha Delphini through magnitude through Beta, through Epsilon, and on for another 3° 45′. Stop at that point and move a degree and a half east, and you should be on NGC 6934. Despite an unimpressive magnitude figure, it will be readily apparent in an 8-inch scope under most conditions. I know NGC 6934 never fails to show itself with ease in the C11. However, even with 11 or 12 inches of aperture, this cluster is not resolvable in the city, appearing as a small grainy spot. In the 11-inch SCT under dark skies, though, considerable resolution is possible with high magnifications, and this little glob also shows itself to be slightly elliptical in shape.

M72

There are Messiers and *then* there are Messiers. There are the M13s and there are the M56s. Aquarius' "other" globular, M72, is definitely in the M56 category. On most evenings, this small, loose globular cluster, glowing weakly at magnitude 9.4, is just on the edge of perception in the 4-inch scope and completely invisible in the Short Tube 80-mm refractor. It usually looks better than M56, but not by much. Applying 12 inches of telescope aperture and dark skies to M72 doesn't make it spectacular, either. It's just ho-hum at best, a dim, loose clump of a star cluster. What I find surprising is that Messier cataloged this one and skipped over a much better globular, NGC 288, nearby in Sculptor. Unlike M72, NGC 288 is an easy and impressive object for smaller telescopes.

If you're up to the M72 "challenge" in the city, the easiest way to find it is probably to start at magnitude 2.95 Alpha Aquarii. Proceed 10° south and west to magnitude 2.9 Beta, and then another 11° 33′ in the same direction to magnitude 3.77 Epsilon Aquarii. M72 is 3° 21′ south-southeast of this star. A pair of magnitude 6 stars separated by a little more than 1° lies just to the west of the cluster. Once you've done M72, be sure to look for NGC 288 as well. It's 1° 45′ from the famous Sculptor Galaxy NGC 253, and about 3° from prominent (when its near culmination) Alpha Sculptoris.

Tonight's Double Star: Mesarthim, Gamma Arietis

Aries, despite its status as a constellation of the zodiac, isn't much to look at. It's an unmemorable pattern of medium bright stars. But one of these anonymous looking stars is a very fine double, Mesarthim. This star is famous in astronomical history as "The First Star of Aries," a name it bore because it was once the closest star to the Vernal Equinox, "The First Point of Aries," before the wobble of the Earth's axis moved the Equinox over into neighboring Pisces.

Both the primary and secondary of this pair are usually said to be of equal brightness, both being given magnitude values of 4.8 by most sources, though the westernmost star does look slightly dimmer to me. The separation of the two components is a generous 8″, which is easy to resolve for most scopes, but not so large as to make the pair less attractive. For me, widely separated doubles are less interesting, looking like nothing more than unassociated field stars in larger scopes. While both of the stars of this pair are white, the "dimmer" component is sometimes (and rather fancifully, if you ask me) called "gray." Mesarthim is not difficult to locate if you can make out the dim hook-shaped pattern of stars that is Aries, lying to the east of Pegasus and Pisces. Mesarthim, Gamma, is the "end" star at the Western terminus of the pattern, closest to Pegasus.

> It's late now, and I'm starting to tire as the stars of Cygnus and Delphinus disappear into the West. But I see Orion 's rising in all his glory. If tomorrow isn't a workday, and if dew and fatigue don't shut me down, maybe I'll observe a few more marvels. I've already almost had a surfeit of wonders. As I stand under the quiet autumn sky amid dead and fallen leaves, the images of these great and mysterious forests of stars linger and will surely haunt my dreams.

Tour 2

Titan and Crab

One of the wonderful things about being an amateur astronomer is the closeness we develop with nature, and particularly with the endless change of the seasons. The average person may scarcely note summer's metamorphosis into fall until the first chill winds blow, but we sky watchers have been anticipating the end of summer long before the Equinox. As August faded into September and October, we saw the Summer Triangle crawl farther and farther into the west and we said a farewell to the myriad wonders and mysteries of the summer sky. This is, I think, a good time to stop and reflect on the unbelievable sights that have paraded across the heavens and our view for the last several months. How many old friends among the summer deep sky objects did you revisit this year? How many new acquaintances did you make? But summer's dead now, and we turn to the east and to the smoky fall stars.

Autumn almost seems like an intermission before the great winter sky show begins. The dull stars of autumn lack the majesty of the brilliant beacons of wintertime—Betelgeuse, Rigel, Capella and, of course, Sirius. Likewise, the DSOs of autumn, the M72s and M1s are not as splashy as the M42s and M35s to come. But this is not to say that there are no spectacular DSOs on-view before winter sets in. For even the most light-pollution-afflicted observers there are some gems waiting in the fall constellations. They are more subtle and delicate than the sights of summer and winter, but I think this fits the contemplative nature of autumn, a time for goodbyes and a drawing-in in preparation for the cold storms of winter.

In spring our focus is definitely on the deep ranges beyond the Milky Way, to the galaxies sprinkled like wildflowers across the great fields of Virgo. In summer, we return to our home galaxy and to the easy pickings of the plane of the Milky Way. The coming of fall again allows us to delve into the void between island universes. Now Pegasus, the great flying horse, sprawls across the heavens. He is peppered with many, many galaxies, but even the brightest of these present high challenges for city astronomers. At best, the Pegasus galaxies appear as barely perceptible fuzz-spots in smaller apertures. But nearby, easily visible in the worst sky glow, is a titan of a galaxy.

I rarely let a clear autumn night pass without taking at least a quick look at M31, the Great Galaxy in Andromeda. I suppose even the greenest deep sky observers have visited this enormous spiral at least once—it's usually one of the first objects new amateur astronomers seek when taking their initial steps outside the Solar System. Unfortunately, most novice observers make their looks at this great wheel of star brief. All too often, I hear novices dismiss M31 as a "bright smudge"—something to be ticked off on the Messier list and nothing more. Extended observation and a bit of study of both the object itself and its vital statistics, however, will reveal subtle glories in almost any object, and M31 is not an exception.

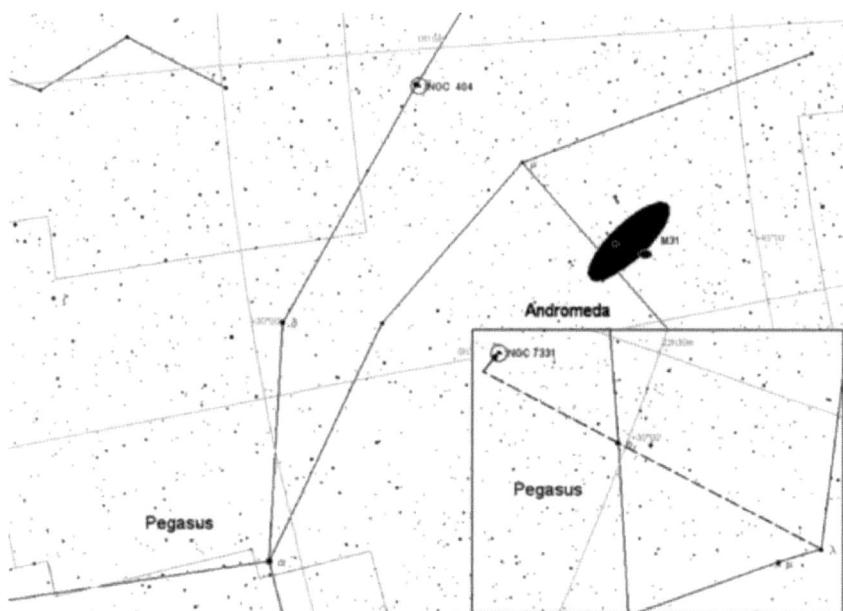

Figure 8.5. Area of the Great Andromeda galaxy.

M31

Summer's journeys have been relatively short hops, usually a mere 20,000 light years or less. Our visit to the more distant of fall's globulars didn't take us much farther. Now we step out into the real dark again. Even the 2.5 million light years to M31 is "backyard" in cosmic terms, but it's far enough that our quest for details in a DSO becomes more difficult—though hardly impossible. Locating M31 is simplicity itself, since it is very large and bright, 178′ × 63′ and magnitude 4.0. Its large size means that its magnitude 4.0 light is very spread out, but it's still bright enough to see with the naked eye from the suburbs, where it appears as a strange nebulous "star." In the city it will be invisible without optical aid, but in binoculars or your finder, it will be easy, and will assume an obviously elongated form.

To find M31, check your star atlas or the chart in Figure 8.5. You'll see that the galaxy lies 1° 21′ west of magnitude 4.5 Nu Andromedae. If Nu is hard or impossible to see in your light-polluted skies, a time honored guide to M31 is the "arrow" formed by the triangle of bright stars composed of Alpha, Beta, and Gamma Cassiopeiae, which should be easy to see in the worst sky glow. Follow this arrow for 15° 16′, and you'll land in the general area of Nu Andromedae. A look through your finder should then reveal M31. When you see the elongated fuzzy of M31 in your finderscope, put it in the crosshairs, insert your lowest power eyepiece into the main scope, and take a look.

How M31 looks at first glance depends on the focal length of your scope more than on its aperture. In a fast, wide-field scope like the Short Tube 80, you'll see a bright, elongated glow with a slightly brighter center. There won't be much, if any, detail visible at first, but M31 will definitely look like a galaxy. With a narrow field scope like an 8-inch SCT, what you'll see is a round glowing ball that represents the nucleus and inner regions of M31. This will be embedded in a fainter haze of nebulosity extending northwest–southeast. You'll have to slew the scope at least one low-power field in each direction to see the full extent of the galaxy visible in the city.

There's no denying that M31 can be disappointing at first blush. I remember drooling over gorgeous long-exposure photographs of this galaxy while I was saving up for my first telescope. Once I'd examined the Moon and Jupiter with my 3-inch Tasco reflector, M31 was the first object I went after. *What a let-down.* The books referred to M31 as "bright," so I expected to be blown away by the sweep of magnificent spiral arms. I found the galaxy easily enough, inserted a low-power eyepiece (I did realize that it would be big), and pressed my hungry eye to the lens. There it was, I'd found it on my first try. I was elated until I started looking in earnest and wondered what was wrong with my new scope.

The blob I was seeing looked nothing like the pictures. All I could make out was a round, fuzzy ball surrounded by a little tenuous haze. I was disappointed enough that I was ready to give up deep sky observing before I'd even gotten started. Luckily, I kept going and found enough "good stuff" to keep me enthused about the deep sky. I was still puzzled by M31, though. Why *didn't* it look like those pictures? I blamed my small scope and forgot about Andromeda.

It took me a couple of years to figure-out why this magnificent galaxy looked so terrible. It wasn't my telescope. No matter what you do or how large a scope you use, M31 will not show spiral structure to the extent that a face-on galaxy like M51 will. It's hard to trace the arms even in photos. This is because of M31's shallow inclination to us. It's viewed almost edge-on, meaning we don't have a good perspective on the arms. Also, like I did the first time I visited here, most novice observers just give this object a quick once-over before moving on. Seeing details in any deep sky object requires more than a 10 second glance. Finally, as always, visually this galaxy will never look exactly like its photos. The eye and the camera are very different sensors. That doesn't mean that M31 always looks *worse* visually, just *different.* The eye actually has an advantage over a camera in that it has a far greater dynamic range than film, meaning it's easier to make out subtle brightness gradations visually.

How do you get beyond the smudge stage with M31? How can you see details in the midst of this fuzzy haze? A good approach is to use a variety of magnifications to scan the whole galaxy, and to use high powers to pull details out of light pollution. After locating M31, start looking for its features with higher power oculars. Begin with perhaps 100×. The trick to making M31 give up detail is patience and perseverance. On many nights in the city, for example, the center of the galaxy is just a round, featureless ball, but on above average evenings at 100× in 6-inch and larger scopes, you may occasionally see M31's "true" nucleus, a tiny star-like point at the center of the fuzzball.

Away from the central region, you'll see the extensive haze that represents the disk and spiral arms of the galaxy. On the best urban evenings, take a look at the Northwest edge of the galaxy. Does this side appear a *little* more sharply delineated than the galaxy's southeast border? If so, you're seeing evidence of one of the dust lanes that

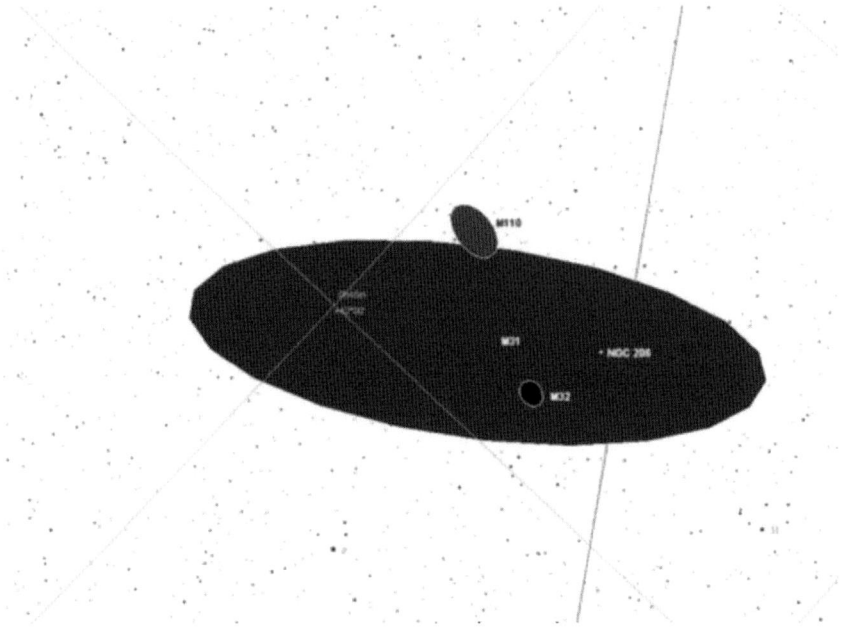

Figure 8.6. Zoomed in on M31.

outline M31's spiral arms. This is not an easy observation to make in the city, but I have done it with a 6-inch reflector when M31 was high in the sky. The secret to seeing the dust lane is to keep trying, especially on transparent late autumn nights after a cold front has passed through.

Dust lane visible? Then keep going. NGC 206, shown in the close-up chart of the M31 area in Figure 8.6, is a huge cluster of giant stars located on the outer fringes of one of M31's spiral arms. It is similar to M24, the Milky Way star cloud in Sagittarius. NGC 206 can be found 30′ west and slightly south of Andromeda's nucleus. Hard? Very. Possible from the city? Yes. When the sky is right, look for a subtle brightening less than 4′ across.

Keep pushing your little urban scope and don't give up too easily on any night. Keep cruising this great galaxy, looking for the star-like nucleus, for NGC 206, and for dust lanes. Be sure to attain as much dark adaptation as possible in your location. Even on the worst nights, though, there is always plenty to see in "Andromeda." In the little 4.25-inch scope on a spectacularly *bad* evening, for example, I found that

> Even under these horrid skies, M31 is quite a sight in the 4.25-inch (48× with a 25-mm Kellner). A large, round nuclear area is visible embedded in very faint haze. M32 is bright and obvious, but M110 is not seen. The view is bad, I guess, but I can help feeling a sense of wonder. I'm looking across nearly three million light years to the home of half a trillion suns!

M32

After the central fuzzy-ball inner regions of M31, the most easily seen object in the area is M32. This is a small satellite galaxy orbiting the center of M31, it is Andromeda's equivalent of our own galaxy's Magellanic Clouds. This round magnitude 9.08, fuzzy lies 25', one medium-power field, southeast of the center of M31. This 8.8' × 6.5' elliptical galaxy, like all its kin, doesn't reveal much in the way of details in any telescope. A higher power eyepiece should at least show that M32's center is brighter than its outlying regions, though. Like most small ellipticals, it looks perfectly round to me even though in reality it's a somewhat elongated Hubble Type E2.

M110

M31 possesses another relatively easy-to-observe satellite galaxy, M110 (a.k.a. NGC 205), which is another elliptical. At magnitude 8.93 and 21.9' by 10.8', its light is more spread out than that of M32, making it surprisingly hard to see from urban observing sites. It is easy to photograph, however, as the sky fogged photo in Plate 44, taken from a substantially light-polluted suburb shows. M110 is the little fuzzy blob to the right of the main galaxy. I have detected its ghostly glow with the 4.25-inch reflector, but only on truly exceptional nights, and it wasn't easy then. Maybe slightly easier than the star cloud, NGC 206. M110 is on the "opposite" side of M31 from M32. Search for it 35' west-northwest of M31's center.

When I've applied high power to M110 with a large aperture scope, I've occasionally thought I've caught fleeting hints of complex detail, as if I'm seeing some grainy clumps or dust spots in the haze outlying its nucleus. Quite likely this is just the result of me straining overly hard to turn up details in what is essentially a featureless elliptical galaxy. Actually, it's not *quite* featureless. Some long-exposure images taken with large telescopes do reveal strange dark patches or lanes in M110, and it has been classified as an E6p, "p" for peculiar. Have I actually detected these dust patches or am I just remembering what I've seen in photos? I'm not sure, but never be deterred from attempting an observation because someone tells you it's "impossible" with your scope and your skies.

M31 and its satellites are located at least 2.5 million light years from Earth (some current estimates put them at 2.8 million light years). The big galaxy is classified as a Hubble Type Sb spiral, since its nucleus and arms are of equal prominence. It is 180,000 light years in diameter, and may contain up to 500 billion stars, making the largest galaxy in the local group, and considerably bigger than the Milky Way. M31 is, unlike the distant red-shifted galaxies in the Virgo Cluster and beyond, approaching the Milky Way, and astronomers believe it is destined to collide and merge with our home spiral in the distant future. The final result of this cosmic collision may be the formation of a monstrous elliptical galaxy that takes the place of our two graceful spirals.

M1

Like M31, M1 held a special attraction for me as a young astronomer. One thing that drew me to it was its "number one" position in Messier's list. Surely, the *first* had to be, if not the best, at least special in *some* way. There were also those beautiful silvery astrophotos of the Crab Nebula. M1 was another of the objects photographed by the 200-inch Hale telescope and made into beautiful black & white prints that circulated widely among the public in the science books and magazines of the late 1950s. In its Palomar print, M1 looks incredibly fascinating. It is a large jagged-edged oval of brightly glowing gas set in a rich star field and overlaid with numerous thin, twisting filaments of gas.

I was also drawn to this nebula, which is a supernova remnant, because I had just been introduced to adult science fiction by Arthur C. Clarke's famous work, "The Star." In this gem of a short story, space travelers come upon the remains of a civilization destroyed by a supernova that turns out to have been seen on Earth as the Christmas Star. I was moved by the story and identified the supernova—unspecified in the tale—with M1's precursor.

Another attractive feature of M1 was its closeness to a bright star, Zeta Tauri, which meant that it was an object I could hope to locate easily in those days when I was learning the sometimes difficult art of star-hopping.

Indeed, M1 is one of a handful of "no-brainer" DSOs when it comes to finding. All that's required is that you be familiar with the bright and famous autumn constellation, Taurus the Bull. The Bull's face, formed by the Hyades star cluster and graced by the tremendous magnitude 1.0 red star, Aldebaran, is a "V" of stars. Each "leg" of this V can be extended for 15° 15′ to a bright star. These two stars, Zeta Tauri and Beta Tauri, are the bull's "horns." The northernmost horn, magnitude 3.0 Zeta Tauri, marks the location of the nebula. Position your scope on Zeta and move just a smidge over 1° south and slightly east and you are *there*. Use at least medium magnification and try to wait for M1 to near the meridian. It's tough in the city, you see.

I was able to land on M1 without much fuss, just as I'd been able to find M31 with ease despite my lack of star hopping experience. That was the only good thing about my encounter with M1. Talk about disappointment. It was even worse than Andromeda. With the 4.25-inch telescope, I could *barely* make out the tiny, dim oval of gray nebulosity seen in my drawing in Figure 8.7. No filaments of glowing gas were to be seen.

I've learned a lot about deep sky observing in the nearly 40 intervening years, but I *still* find M1 to be rather difficult and bland from light-polluted areas. At an integrated magnitude of 8.4 and 8′ across its major axis, M1 has a respectable surface brightness of 11, but looks dimmer, much dimmer than that to me. It is *intrinsically* faint, and, like nebulae of all types, takes a real beating from light pollution. I recently came back to M1 with the 4.25-inch reflector, the very same telescope I'd used on it as a young amateur, to see if I'd been too hasty or hadn't known what I was doing back in the old days:

> If you're observing the Crab from the city with a 4.25-inch telescope, you have to be satisfied just to say you've seen it. I didn't have much trouble picking it up, but only with averted vision. I can't see it with direct vision in this telescope on this evening. The only distinguishing feature beyond its small, pale gray oval is that it is obviously elongated, and even that is not an easy observation to make. A few dim stars are scattered across this lonely field.

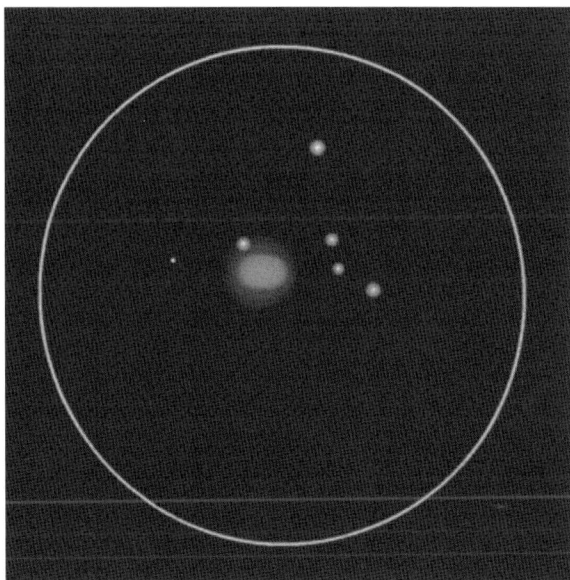

Figure 8.7. M1, the Crab Nebula in a small scope.

Can you do better? Possibly. How? Aperture. I don't notice *much* change in M1 before 12.5 inches. Until you get there, M1 becomes easier to see with each jump in light-gathering power, but remains a featureless oval. In 12.5-inch scopes, though, a couple of things become apparent. With sufficient magnification, the Crab morphs from an oval to something that looks more like the lightning bolt shape that's seen in the image in Plate 45. This aperture class also begins to reveal that the body of the Crab is not smooth, but has a wispy, filamentous character around the edges.

To see the tendrils weaving in and out across the main body of the nebula takes a lot more aperture in the city than 12.5-inch scopes and judicious application of an OIII filter. I have seen the filaments and streamers with my 12.5-inch telescope under dark skies, but the smallest instrument that has revealed them to me in the city has been a 24-inch Dobsonian. Even then, an OIII and some careful observing were required. The Crab looked nothing like its image, but with the OIII in place and at high power, I could see where at least one filament crosses the Crab's body and branches into two. Unfortunately, the OIII suppresses M1's main nebulosity, so getting a true idea of the object, even with a large scope, required me to switch the filter in and out, and increase magnification to over 500×.

Despite the frustrations Old Crabby poses to visual observers, he is still an interesting stop. Perhaps more interesting historically than at the eyepiece, however. Despite its position in the number one spot in his catalog, this object, like surprisingly many of the others, was not *discovered* by Messier. The first person to have seen, or at least recorded it, was the English amateur astronomer John Beavis in 1731. Messier appears not to have heard of Beavis' observation and may have discovered it independently in 1758. M1 probably *was* the object that spurred Messier to begin his list with the initial intent of cataloging comet-like objects—it certainly looks like a dim comet in a small instrument. Messier's goal, at first, anyway, was to help his fellow comet observers avoid wasting time with funny comet-like patches that didn't move. The first observer to get a *good* look at M1, however, was Lord Rosse, who made a drawing of it in 1844 from his Irish castle. It was Rosse, as a matter of fact, who christened this supernova remnant "The Crab." The filaments, which were visible in his "small" scope, a 36-inch, reminded him of the claws and legs of a horseshoe crab. His eyepiece drawing looks more like a pineapple than a crab to me, but the name stuck.

Much of M1's fame among professional astronomers is due to the remains of its progenitor, a rapidly spinning neutron star, the battered core that is all that's left of the giant sun that died in the supernova. Since it shines dimly at magnitude 16 in the middle of the Crab's comparatively bright nebulosity, city observers, even those with large scopes, don't have a prayer of spotting it, I'm afraid. This neutron star is composed of "degenerate" matter that has been so compressed by the supernova explosion that the whole star is really best thought of as a single giant neutron. As it spins, the interaction of the star with its magnetic field causes a radio beam to be emitted and rotate lighthouse fashion across the stars. This "pulsar" was one of the first stars to be identified, not long after the initial discovery of these of radio wave-emitting neutron stars by Cambridge University's Jocelyn Bell in 1967.

My romantic association of the Crab with the Christmas star is demonstrably false, as studies have shown that it is undoubtedly the remains of the supernova observed by Chinese astronomers beginning on July 4, 1054. It was also seen by native peoples in the Americas, but seems not to have been noticed in Europe (maybe it was a cloudy summer). For once, the distance to a DSO is well known, with M1 having been determined to be 6300 light years distant. This makes its constantly expanding cloud currently 10 light years in diameter.

Other Area Attractions

NGC 404

At a relatively dim magnitude of 11.23 and a relatively large size of 3.4′ across its major axis, this oval-shaped galaxy should be a *challenge*, but not the *huge challenge* it often is for small urban telescopes. The difficulty lies in its close proximity to the gloriously bright star Beta Andromedae, which is only 6′ 46″ away, shining at magnitude 2.0. The galaxy is easy to find, but rarely easy to see in the glare of its "companion" star. NGC 404 is, however, regularly visible in an 8-inch scope, and I have seen it from my city observing sites with a 6-inch scope at times. The secret to conquering it is to use a high enough magnification to enable you to put Beta Andromedae just outside the field and NGC 404 toward the field center. Try 150× to begin with. Some observers claim a broad-band light-pollution reduction filter can dim the star a little while preserving the galaxy enough to make this an easier observation.

NGC 7331

Moving into western Pegasus, to the prominent triangle of Eta, Mu, and Beta Pegasi, the stars that make up the Flying Horse's forelegs, we find NGC 7331, a magnitude 10.33 Hubble Sab spiral that is often referred to as "Andromeda Junior." To find it, draw a line from Mu, through Eta, and into space for another 4° 20'. NGC 7331 lies at the terminus of this line. There are no bright stars in the area, so be careful and refer to a detailed star chart generated with *Cartes du Ciel* or *Skytools 2*. Its "Andromeda Junior" nickname reflects the fact that its inclination to us is similar to that of M31, and, in small to medium-sized scopes, it does look like a perfect miniature of the Great Andromeda Galaxy.

Under dark skies, scopes 12 inches in aperture and larger reveal considerably looser spiral structure in NGC 7331 than is seen in M31, making one spiral arm stand-out dramatically from the disk, but this feature is invisible in the city. In light pollution the galaxy is an object most suited for scopes 8 inches and larger in aperture. In an 8-inch scope on a dry night, NGC 7331 is visible as a clearly elongated fuzz-spot that shows off a bright core and not much else.

You'll occasionally hear NGC 7331 referred to among amateur astronomers as a member of the "Deer Lick Group" of galaxies. There are many dim NGC galaxies occupying the same field as this object, and I suppose they represent deer licking at the cosmic salt of NGC 7331. These tiny cosmic lint balls are at daunting magnitudes dimmer than 14 and are likely invisible in the city in any but the largest scopes. I have seen them with some effort in my C11 under dark skies, but never from any site near the city.

Just 29' south-southeast of the Deer Lick is the legendary galaxy group Stephan's Quintet. Sadly for us, these five tiny galaxies are dim and close together and will probably take something on the order of a 20-inch instrument on an exceptional night in the city to see. Even in the country, they, like the Deer, are often barely detectable as minute magnitude 14 and dimmer smudges in my 11-inch Schmidt Cassegrain.

Tonight's Double Star: Almaak, Gamma Andromedae

Almaak is another almost perfect double star for the urban observer. It features a combined magnitude of 2.3, making it the third most luminous star in its constellation, and its separation is a wide 10", meaning that it's easy for most scopes to resolve despite the secondary star's relatively dim magnitude of 5.5. The magnitude 2.3 primary is a burnished gold; while the secondary is a deep blue, hearkening back to summer's magnificent Albireo.

Gamma Andromedae is not just a double star, it's a triple. The blue secondary star has a companion of its own. Unfortunately, this companion lies a scant 0.4" from its parent star. It's not overly dim at magnitude 6.3, but the extreme closeness means you'll need steady seeing and big aperture and luck to see it. Almaak is easily found, being the end star on the easternmost of the two chains of stars that form Andromeda's main figure.

> After a long, satisfying view of M31, I find it difficult to pull myself back across the light years and return to the now-insignificant problems and worries of a minor planet orbiting a puny G2 star on the outskirts of an average spiral galaxy. It's time to pack my wonderful telescope away for the night, but I know that when earthly problems and worries again loom large, the wondrous voyages of amateur astronomy will bring them back into proper perspective.

Tour 3

The Cassiopeia Clusters

I seem to do some of my best astronomy thinking in the early morning—or even while I'm asleep. So, I wasn't surprised to awaken one day with the idea for a project I could execute from my light-polluted home in my city's Garden (historic) District. I would observe as many open clusters in Cassiopeia as a 12.5-inch telescope would show me. I've always enjoyed drifting through the Milky Way in the Cassiopeia area, but the last time I'd taken a detailed look at this part of the sky—many years previously—my only telescope had been my 4.25-inch $f/11$ reflector. I was interested in seeing what a larger telescope could pull out of my bright skies.

I was also curious as to whether the clusters I'd observed in the past would look much different through a larger instrument. I know I've stressed the practicality of smaller telescopes for urban observers, but, as I mentioned in Part I of this book, a 12-inch aperture Dobsonian, is not a huge hassle to set up. If you have a private ground-level observing area, an instrument in this class may be just the telescope for you. There is no question that my "big" dob allows me to penetrate my site's depressing sodium streetlight haze more deeply than I can with a smaller instrument.

But don't hesitate to undertake this tour even if you're equipped with a considerably smaller telescope. A 6-inch reflector or 4-inch refractor may show a little less than what I saw if your skies are as bad as mine, but may show you *more* if your sky glow is less pronounced—or your eyes or observing skills are better than mine.

A look at *Sky Atlas 2000* and a session with *Skytools 2* and *Cartes du Ciel* revealed there'd be approximately 40 clusters in the area that would be bright enough to provide interesting targets for my 12.5-inch scope. A little more narrowing-down to the "best of the best" (I tended to eliminate large and sparse groups) left me with an observing list containing 17 objects. After some time spent familiarizing myself with the locations and characteristics of my of destinations, all that remained was to wait for clear skies, which took a while, since we were experiencing one of our typically stormy Gulf of Mexico Coast Octobers.

When that rare beautiful night finally arrived, I was more than ready to begin my tour of the Celestial Queen. I didn't expect to finish in one evening, and I hope you don't either if you decide to follow me. I wasn't interested in merely ticking objects off a list; I wanted to spend some time in these stellar nurseries, to try to absorb some true sense of them.

Most of these galactic clusters are easy to find, being located near Cassiopeia's "W" asterism. If you run into trouble with the objects situated away from the main star pattern, make sure your finder is up to snuff. Throughout this book I've mentioned 50-mm finders again and again. A 50-mm finderscope reveals enough stars to make locating dim objects easy. If your finder is smaller than this, you're making things unduly hard on yourself. Actually, I'm toying with the idea of adapting my 80-mm $f/5$ Short Tube Refractor as a "super finder." As long as it offers a wide enough field, your finder really cannot have too much aperture.

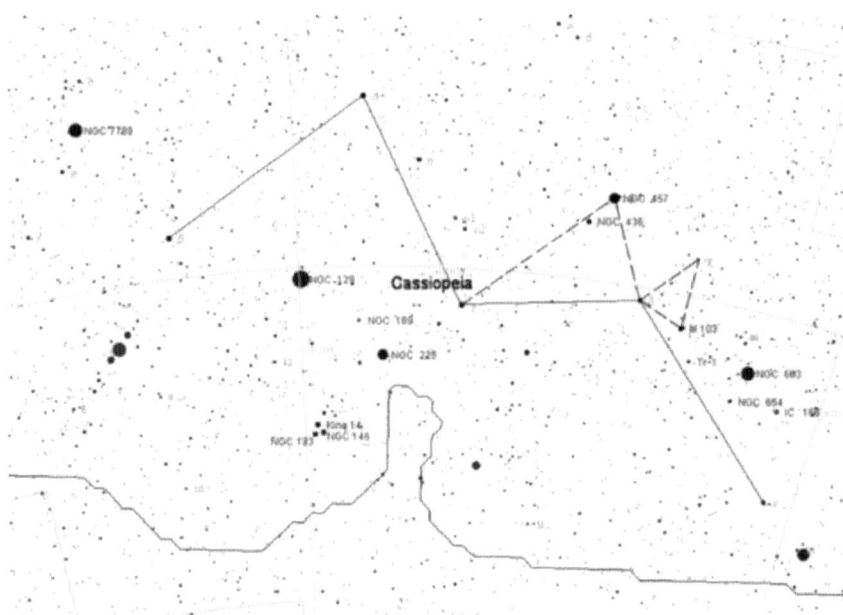

Figure 8.8. Cassiopeia, the Star Queen.

It's a crisp and clear October evening, the kind that is so beautiful you can't stand to stay inside. An occasional breeze brings a hint of coming winter's chill. All around is the cozy scent of burning leaves, reminding us of countless autumns past. Overhead, the stars of fall glimmer. In the far northeast Cassiopeia, Queen of the Sky, rides ever higher as the great spinning wheel of heaven keeps on rolling. This would be a fine, fine night for naked eye or binocular stargazing, even here under city lights, but my wonderful telescope stands ready. With a detailed star atlas or Figure 8.8 in hand, let's set off to see what can be seen in the domain of the Star Queen.

NGC 457

I suppose that the logical way to conduct a tour of Cassiopeia is to start at one end of her "W" and work your way down the constellation stick-figure, but my wife, Dorothy, was very interested in getting a look at NGC 457, the E.T. (or Owl) cluster, so I started there, working my way northeast from his location. ET is compact and bright at 13′ across his long axis and magnitude 7.0, and fits the field of a medium/low-power eyepiece very comfortably. This cluster is spectacular in even tiny instruments.

E.T. is located 2° South of Delta Cassiopeiae, forming an isosceles triangle with Delta and Gamma, and is visible with ease in a finder, appearing as a small "line"

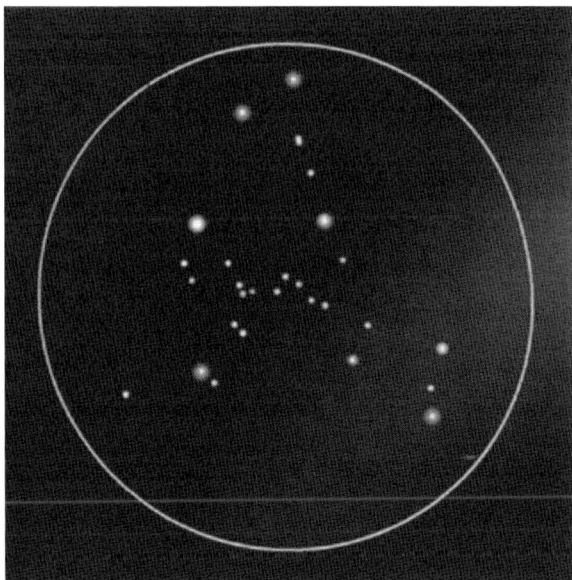

Figure 8.9. E.T., aka "The Owl Cluster."

of stars at low magnifications. When you have NGC 457 in the field of your main scope, play "connect the dots" with the suns in this sparkling cluster to form a stick figure. Soon enough, you'll begin to see a little extraterrestrial, hand raised in greeting (Figure 8.9 is my attempt to sketch the little guy). One of his eyes stands out brightly against the rest of the cluster. This is magnitude 6.99 Phi Cassiopeiae. While most of the rest of the stars in your eyepiece do belong to NGC 457, Phi, the most brilliant star in the field, is actually a foreground object not physically associated with the group. The view through my 12.5″ was amazingly good, but even a Short Tube 80 reveals much of NGC 457's lustrous body. About 40–45 cluster members are easily visible in the 12.5-inch scope. A red star in ET's left armpit is very distinctive.

NGC 436

Our next destination is NGC 436, which is located only 49′, maybe one low-power field depending on your scope and eyepiece, to the northwest of NGC 457, back in the direction of Gamma Cassiopeiae. In the 12.5-inch scope, this 6′ diameter, magnitude 8.8 open cluster is similar in appearance to one of Cassiopeia's most spectacular objects, M103, as seen by a 6-inch telescope (i.e., very beautiful). Approximately 25 magnitude 11 and fainter stars are easily seen. The cluster is quite compact, and looked best at 90× in the 12.5-inch scope with a wide-field Konig design eyepiece.

M103

Continuing on, move 3° to the north-northeast to find the next port of call, M103. It forms a near equilateral triangle with bright Chi and Delta Cassiopeiae. Lovely! When you've got this group of newborns in the eyepiece, you'll see a handful of brilliant blue gems with a striking orange star positioned near the center of this magnitude 7.4 cluster. At least 25 stars were visible across the cluster's 6.0′ extent in my light-polluted skies, with the stars arranged a vaguely triangular shape. With 12.5 inches of aperture, M103 almost seems to display a 3-D effect (I again used 90× to good effect).

Trumpler 1

Moving northeast for another 41′ brings the relatively bland Trumpler 1 into view. It was small, less than 5′ in diameter in my 12.5-inch scope at medium magnification, but relatively bright at magnitude 8.1, so it was easy to locate and distinguish from the area's star-rich background. I would guess it could be a little dim in an 8-inch scope, though. 10 stars are arranged in a square asterism that defines the cluster. "TR1" is worth a visit because of one prominent red star in the middle of the grouping. Due to its small size, this cluster was best seen in a 12-mm Nagler eyepiece yielding 127×, and a little more power would probably have improved it even more.

NGC 654

Continue on a northeasterly course (turning a little bit more to the east, now) and you'll arrive at another stopping place along the star trail, NGC 654, a little over 1° away from TR1. Like Trumpler 1, NGC 654 is small, 5′ in size, but it is considerably brighter than the previous cluster at an integrated magnitude of 6.5. In a medium-aperture scope, 25 stars are visible in this sassy group. One striking yellow sun stands out from the crowd at 127× in the 12-mm Nagler eyepiece's spaceship-porthole field. A pleasing view indeed.

NGC 663

From NGC 654, change compass heading and move 1° southeast for the next stop-over, NGC 663, a big 16′ wide magnitude 7.1 association. To pin it down, move 40′ to the east of the midpoint of a line drawn between Epsilon and Delta Cassiopeiae. NGC 663 is beautiful in any scope on any respectable city night. Forty bright luminaries and many fainter cluster members are visible in 8-inch and larger telescopes. A 27-mm Erfle eyepiece (56×) did a good job on this one for me with the 12.5-inch scope.

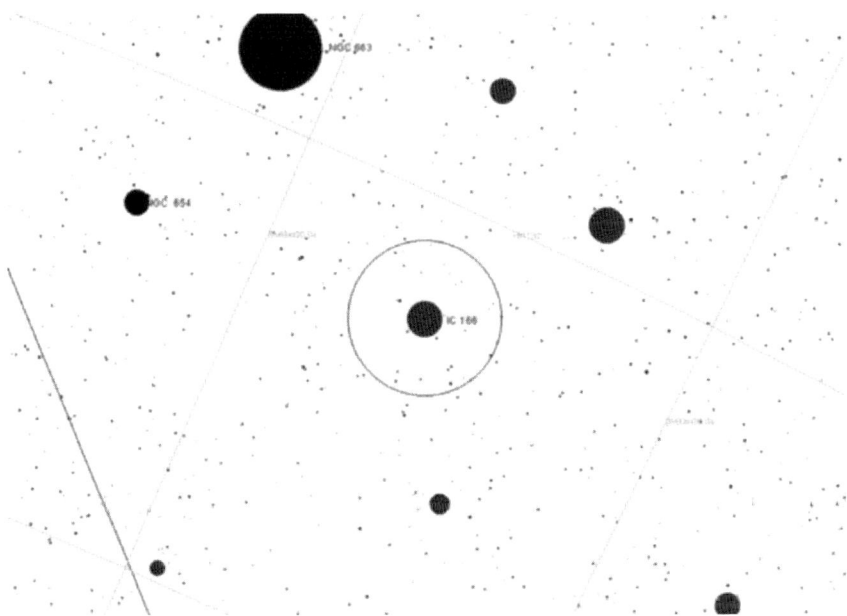

Figure 8.10. Detailed finder chart for IC 166.

IC 166

Switching directions again and heading back northeast for 52′ takes you to the area of the only object in this tour that I found at all difficult with the 12.5-inch Dobsonian. It's small at 5′, but challengingly dim at magnitude11.7, and I had a rather hard time locating IC 166. It is extremely easy to miss in the midst of Cassiopeia's busy star fields (most of the IC clusters were discovered photographically as vague clumpings in rich star fields on photographic plates). I was finally able track it down, but only by using a computer generated small-area finder chart like the one in Figure 8.10, which shows field stars down to approximately magnitude 13. Once I found it, this cluster was hardly spectacular, with a few faint members winking in and out. IC 166 is reputed to contain nebulosity, but I certainly didn't see any. This cluster may look better and be much easier to find if your skies are darker than mine, so don't let me scare you off from it. This object is not shown in *Sky Atlas 2000.*

M52

You have now arrived at the eastern terminus of tonight's guided tour, so take a breath and hop all the way to the western end of the Queen's W to spectacular M52 (Plate 46). Draw a line through Alpha to Beta Cassiopeiae and on for another 6°, carefully position your scope on this spot, and M52 should literally hop out of the eyepiece at you. It is readily apparent even in a large-aperture finder under most city conditions. Another way of looking at its position is that it forms an isosceles triangle with Zeta and Iota Cephei in the neighboring constellation, Cepheus.

M52 is a spectacular object, as you'd guess from its membership in the Messier catalog. Measuring 13' across and glowing strongly at magnitude 6.9, it is very compact and pretty, and is made most beautiful by the presence of a prominent ruby nestled among its sapphires. This bright star located on the southwestern edge of the cluster is, as is often the case with these stellar stand-outs, not an actual member of the group. M52 is very rich, and more than 50 cluster members were readily visible in my heavily light-polluted skies with the Dobsonian. If ever the phrase "like diamond dust on black velvet" described a cluster, it is M52.

NGC 7789

Another large hop, 6° 25' this time, to the south-southeast brings up a group I thought was even nicer than M52, and which is rivaled only by NGC 457 for the title of "most beautiful star cluster in Cassiopeia." NGC 7789 (Plate 47) is another bright and easy catch at 16' in extent and magnitude 6.7. Look for it halfway between magnitude 5 Sigma Cassiopeiae and magnitude 4.5 Rho. Just outstanding! A showpiece even when dimmed by heavy sky glow. Very rich, it resembles a loose globular like M71 in Sagitta. A goodly number of magnitude 10 stars and many, many magnitude 12 and fainter cluster members we visible in my medium-aperture telescope.

NGC 129

At 5° 30' northeast of NGC 7789 on the other side of Cassiopeia's W is NGC 129, which lies near the center of a line drawn between Beta and Gamma Cassiopeiae. 21' diameter and glowing at magnitude 6.5, you'd think this one would look much like NGC 7789, but it's nowhere near as rich or condensed. It tends to the large and sparse, but is made impressive by the inclusion of bright DL Cassiopeiae in the field, a sixth magnitude star that is not a cluster member. The cluster was attractive in my 12.5-inch scope with a 27-mm Erfle eyepiece at 56×, but 90× brought out some fainter cluster members. Use a fairly high-power wide-field eyepiece on this one, if possible.

NGC 189

Now, hop 1° 31' northeast to magnitude 8.8 NGC 189. This open cluster is located 1° south of the midpoint of a line drawn between magnitude 5.5 12 Cassiopeia and Gamma Cass. NGC 189, a little thing, less than 5' in diameter, is distinguishable from the starry background, but there's not much there, really. About 10 stars, of magnitude 11 and dimmer were detectable in the 12.5-inch scope.

NGC 225

Another degree away on this northern path is the much nicer NGC 225, which is 50'
to the north of NGC 189. What a very nice surprise this little cluster was. A big 12',
magnitude 7.0 nest of suns, it really breaks through the light pollution in medium-
aperture scopes. Here, 15 stars are arranged in a pattern most observers call a "W". To
me, though, it looked just like a tiny, perfect Sagittarius Teapot.

NGC 133, NGC 146, and King 14

2° west of NGC 225 is the end of the road for this evening's ramble, the area of three
small clusters I thought were both pretty and unique: NGC 133 (mag 9.4, 7'), NGC 146
(mag 9.1, 7'), and KING 14 (mag 8.5, 7'). These open clusters are located near Kappa
Cassiopeiae in an area 30' northwest of the star. This is a rich locale, and, while the
three clusters do stand-out fairly well from the background star fields, it is somewhat
difficult to decide where each cluster begins and ends. A very nice sight since all three
objects are visible in one low-power eyepiece field (they are all about 12–15' apart).
Use your lowest powered, widest field ocular. There is some feeling, by the way, that
NGC 133 may be an asterism rather than an actual star cluster.

Tonight's Double Star: Achird, Eta Cassiopeiae

Achird, Eta Cassiopeiae, is an intensely lovely double star. The brighter of the pair,
a golden yellow magnitude 3.4 sun, is accompanied by a considerably dimmer red
companion, which glows balefully at magnitude 7.5. The separation between the two
is a respectable 12", but the large difference in magnitudes makes this one a little
tougher than you'd think for the smallest scopes. I see the secondary as a deep red, but
many observers swear it's *purple*. This is likely due to a color-contrast effect similar to
the one that turn's mighty ruby Antares' dim white companion a shocking emerald.

Locating Achird is easy thanks to its proximity to Cassiopeia's "W" asterism. It is
about one-third of the way along a line drawn between two of the W stars, Alpha
Cassiopeiae and Gamma Cassiopeiae. Go about 1/3 of the way along this line from
Alpha toward Gamma and stop, and you'll immediately notice Eta shining bravely
about half a degree to the east-southeast.

If you think Achird is tough to split with your small telescope, come back in a
decade or two and it will be a little easier. This is an actual binary pair; the stars are
orbiting a common center of gravity, and their separation is increasing slowly due to
the companion's elliptical orbit. Unlike Achird, many of the double stars we observe
are not "true" binaries; many are simply optical illusions created by our viewpoint.
Sometimes two members of a pair are just along our line of sight and are not physically
connected. The status of a double as a true binary star or "merely" an optical double

must be determined by long observation of the stars' motions or by spectroscopic study.

Was I a little tired of open star clusters when I finished this observing project? Maybe a little. But my sense of wonder was restored when I stopped and thought about the implications of what I had seen. Using an inexpensive medium-aperture telescope to scan this medium-size constellation from my bright backyard, I'd visited open cluster after open cluster—and barely grazed the surface of what can be seen there. The huge nests of newborn stars I saw on this night represent one tiny area of the Milky Way. Now do you begin to realize how big our galaxy is? I did, and was awed.

Tour 4

Deep Water Constellations

Late autumn always seems magical to me, and not just because the year-end holidays are on their way. Some of my fondest observing memories are of clear boyhood November nights under the stars. Back in the 1960s, before the unchecked and mutant growth of strip malls and automobile dealerships, a young observer with a *very* modest telescope could voyage almost endlessly thorough the deep sky. The combination of crisp, cold nights, the almost unbelievably bright stars of the Winter Milky Way, and my beloved 4.25-inch Newtonian made for many unforgettable observing runs, the memories of which are still dear to me today.

Face south as winter comes in, and you'll see the autumn constellations on perfect display. These "watery" star figures, Aquarius, Capricornus, and Cetus, about to give way to the winter star figures, are perfectly placed for early evening viewing. There are some subtle and beautiful and often ignored deep sky objects in the area, and we'll visit a pair of the water area's prime constellations, Capricornus and Cetus tonight, but let's start with a star pattern that's located just to their South and *not* associated with water.

NGC 253

I hesitate to use the phrase "off-the-beaten-path" when talking about this constellation, since just about every observing guide at least mentions the wonders of this area, but a Northern Hemisphere bias on the part of some authors gives Sculptor the short shrift. Yes, Sculptor *is* located at a southerly declination, and is, I guess, just about at the limit of what's practical for many northern U.S. and European observers to observe. However, even from northerly latitudes, there's a true showpiece visible here, a galaxy so wonderful I could easily spend an entire night admiring it. Get your telescope ready and come along with me—you're in for an incredible experience.

I suppose there's no use trying to convince you that M31 is *not* the greatest galaxy visible from the Northern Hemisphere, but I'm here to tell you that Sculptor's NGC 253 is a close second. This huge spiral, located near the south galactic pole, was a revelation for me the first time I located it, since it was not only bright, but showed considerable details in my 4.25-inch scope—something galaxies are simply not supposed to do when viewed by small scopes from light-polluted neighborhoods.

How do you find NGC 253? It's easy. All you'll need is a clear southern horizon and a decent star atlas or Figure 8.11. By mid evening on a cold, clear November night, Sculptor is transiting the meridian and NGC 253 is well placed for viewing. Though Sculptor is a somewhat faint and lackluster constellation, it's fairly easy to spot, since there's not much else in the area. Sculptor's stars aren't overly prominent, but they stand out well in their isolation. Predictably, the constellation doesn't look anything like a sculptor or a sculptor's tool. About the only thing I can compare it to is

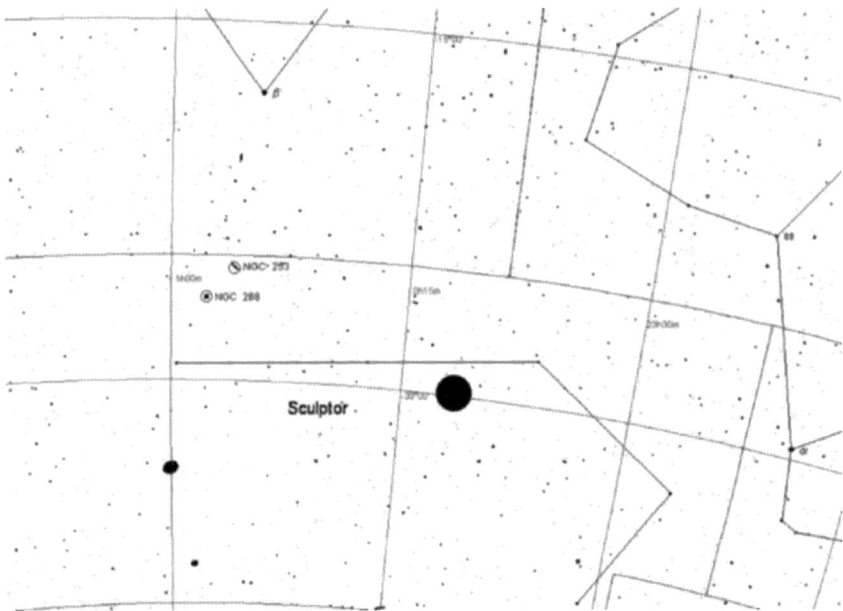

Figure 8.11. Sculptor and the area of the South Galactic pole.

a backwards letter 'j' lying on its side. A good guide to Sculptor (and NGC 253) is the magnitude 2.0 star Beta Ceti. NGC 253 itself is big at 27.7' × 6.8', but it's also bright for a galaxy at an impressive integrated magnitude of 7.72, so looking for it should cause a minimum of heartburn.

Draw a line almost due south from Beta Ceti for 12° and you'll come upon the magnitude 4.3 star, Alpha Sculptoris, which is easily visible in a finder in the barren reaches near the south galactic pole (which lies a few degrees northwest of Alpha). A little over halfway along this line, about 7° 15' from Beta Ceti toward Alpha, turn and go 1° 20' west of the line, and you'll find our quarry, as shown in the chart. A triangle of sixth and seventh magnitude stars just east of the galaxy also provides a good guide. You may be able to pick NGC 253 up in a 50-mm or larger finder on special nights. The sky-cleansing effects of passing cold fronts will help.

Were you surprised by your first look at NGC 253? I know I was. The fact that the galaxy is very large and fairly low in the sky even from my southerly latitude led me to expect very little. But I was *amazed* when I found this DSO. NGC 253 *is* large, but it's also bright. In my little 4.25-inch ƒ/11 reflector, the galaxy's mottled appearance, tremendous dark lanes and bright patches were easily discernable. Once in a while, I even felt a hint of this giant's subtle spiral structure Averted *imagination?* Even in those days, the neighborhood where I grew up didn't feature the skies of the Texas Star Party.

I've viewed this DSO many times since, but it's always a treat and a reminder of why I got started in amateur astronomy to begin with: to see and try to understand the marvelous and mysterious. A recent observation with the same 4.25-inch Newtonian from my now badly light-polluted city showed that the little scope could still deliver a lot of galaxy from today's worse skies:

> I passed over NGC 253 several times before I realized that I needed to push the magnification as low as it would go—for a change—-to help make this wonder stand out. I had to get enough dark space around the galaxy to be able to pop it out of the sky. When I had it in a 32-mm Plossl, it appeared as a smoky cigar shape. Extended observation brought out hints of dark patches.

This fine galaxy bears the distinction of being discovered by one of our most dedicated—and most overlooked—amateur astronomers, Caroline Herschel, the sister of William. Ms Herschel discovered this jewel during one of her tireless comet searches in 1783. Though Caroline is understandably overshadowed by her remarkable brother, her glorious and productive career makes a story I'd like to see told more often.

The technical details of NGC 253 paint a picture of a large and fairly nearby Sc spiral. Sc is a provisional designation, as the galaxy's almost edge-on inclination makes it hard to be certain of its exact Hubble type. NGC 253 is the brightest member of the Sculptor Group of galaxies, which is estimated to lie about 8.1 million light years away (or slightly more distant than the M81/82 group). This is a big and healthy spiral with a mass of around 150 billion Suns. At times, its mottled appearance, well seen in Plate 48, makes it superficially resemble "disturbed" galaxies like as M82, but a look at a long exposure photo makes clear that this is a *dusty*, but otherwise normal and untroubled member of the cosmic zoo.

NGC 288

Did you hop over to NGC 288 from M72 earlier? If not, let's go there now. This magnitude 8.1, 13.8' diameter globular star cluster lies south and east of NGC 253, back on the line from Beta Ceti to Alpha Sculptoris, 1° 45' from NGC 253, or only a couple of low-power fields away. Slew southeast and this star cluster should appear in your field almost effortlessly. How it will appear when you've got it will depend both on the condition of your skies and your *latitude*. From home, 30° north, I am able to pick stars out of NGC 288 with little difficulty with the 8-inch f/5, though I did have to run the magnification up to 200× to darken the background sufficiently. From southern latitudes, this cluster must be magnificent indeed.

M77

Sculptor conquered, let's dive into the watery realm of Cetus the Sea Monster. Cetus is a huge constellation and his star pattern is not always easy for me to make out in the southern sky-ocean. Not only are my city skies badly light polluted, some of the worst sky glow is to the south—the Gulf of Mexico is filled with brilliantly lit oil rigs. Unfortunately for me, there are many more interesting objects

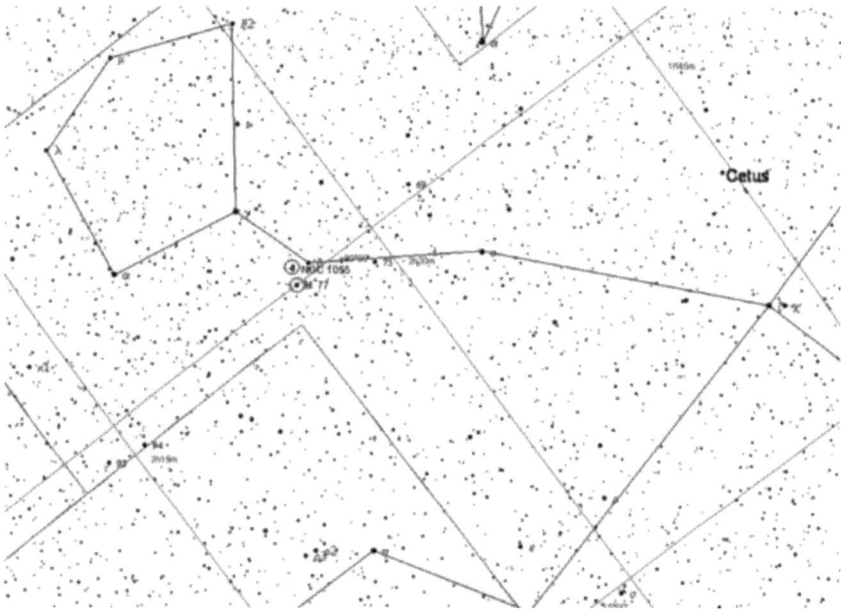

Figure 8.12. Chart of Cetus the whale—sea monster.

along the southern horizon than there are along my comparatively dark northern horizon.

The whole area from Cetus through Sculptor is clogged with galaxies. This is the southern celestial hemisphere equivalent of the Coma Berenices area. Here, you're looking out of the galaxy in the direction of the South Galactic Pole rather than the north galactic pole. Unfortunately, there's no equivalent of the bright Virgo cluster here. There are many distant giants lurking, but most of them are invisible in small city telescopes. M77 is one of the few exceptions. It's not only prominent in gray skies, it's easily located. Take a good look at Figure 8.12 and position your scope on magnitude 4.08 Delta Ceti, to the west of Cetus' "head." Then, slowly move the telescope 1° south and just slightly east, keeping an eye out for a suspiciously fuzzy star entering the field.

M77 is a real standout among the dim southern spirals, a galaxy so prominent and easy that it bears a Messier catalog number. This face-on Sb spiral is not intensely bright at magnitude at 9.64, but it's not overly large at 7.0′ × 5.9′, so even if your scope won't show anything else in this area other than stars you *will* see M77. I found it with ease in my ETX 60-mm refractor, which showed it as a slightly fuzzy "star." In a 3–4-inch telescope M77 is unmistakably nonstellar, consisting of a bright core and a small disk of faint haze. Visually or photographically, even in images taken with sensitive CCD cameras like Plate 49, M77's tightly wound spiral arms are hard to make out. What's amazing is the galaxy's overwhelmingly bright core. M77 is classified as a "peculiar" Seyfert galaxy with an active nucleus like Canes Venatici's M94. Bottom line? A black hole is probably feeding on unlucky stars at M77's center.

NGC 1055

While you're in the area, try for a "bonus" galaxy, NGC 1055, which is situated half a degree northwest of M77. An edge-on spiral at a visual magnitude of 10.60, it is not too much dimmer than M77, but is considerably harder to spot due to its lack of a bright central region. On nights when the southern horizon is not too bright with light pollution, I can see this one with some difficulty with the C11 SCT.

M30

M77 is a rather strange looking galaxy and M30 is an odd looking globular. At a magnitude of 7.5 and a diameter of 11', this globular star cluster is Capricornus' only claim to fame other than its position among the zodiacal constellations. With the exception of some frighteningly dim galaxies, there's not a whole lot else to see in this area, even for large amateur scopes from dark sites, but M30 is a huge treat.

I am able to find this cluster with ease in almost any instrument by using magnitude 3.77 Zeta Capricorni, located on the Eastern side of Capricornus' kite/triangle shape (the Seagoat looks even less like its namesake than most constellation figures). From Zeta, slew 3° 20' east and slightly north. The cluster shows up well, and is somewhat resolvable in light pollution with 6-inch telescopes. This is demonstrated in my 6-inch

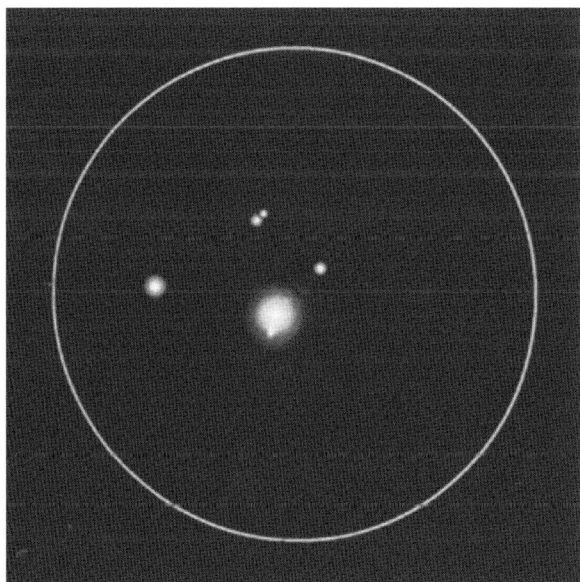

Figure 8.13. M30, the lop-sided globular cluster.

scope drawing in Figure 8.13, which shows some stars despite this glob's fairly tight rating of V (5) on the Shapley–Sawyer scale. I tried to capture the cluster's offbeat, off-balance appearance in the drawing. What makes M30 look odd is the three "lanes" of stars on its northern edge. These give it a strange off-center appearance. Even when not fully resolved, these star chains lend a weird, clumpy, half-circle aspect to this fine globular.

M30 is one of Charles Messier's original discoveries; he saw it for the first time in 1764. At 26,000 light years away, it is approximately 90 light years in diameter, assuming that that distance figure is correct, which it probably isn't. The brightest stars in M30, bloated red giants at the ends of their lives, are at magnitude 12.0, enabling fairly small scopes to retrieve some suns from its misty globe despite streetlights. Although it's a fairly normal looking globular, despite the odd star chains mentioned above, M30 is one of those clusters thought by astronomers to have undergone "core collapse." Although the cluster's center looks normal compared to M15's blazing center, it is believed that at least half of M30's huge mass is concentrated in an area a "mere" 20 light years in diameter. The "why" for this phenomenon is still unknown.

Tonight's Double Star: Kaffaljidhma, Gamma Ceti

Gamma Ceti is a lovely if somewhat challenging double residing back in the dim water constellation, Cetus. While sometimes resolvable with a 4-inch scope under excellent conditions, it can be a trial for an 8-inch scope under average or poor seeing to show as two separate stars. This is because its primary and secondary are of unequal brightness at magnitudes 3.5 and 7.3, respectively, and quite close at a 2.8″ separation. Nevertheless, this is one of the most popular "harder" doubles in the sky, maybe because observers are attracted by the contrasting colors of the primary and secondary. The main star is a very pale yellow, while the companion is an icy blue.

Gamma is no harder to find than anything else in dim Cetus, and the fact that it's one of the "stick figure" pattern stars helps. Gamma is the star that connects Cetus' "head," the northern part of the constellation, to the rest of the Sea Monster's sprawling body.

> NGC 253 is wonderful in the city, but my most cherished memory of this object comes from one year's Mid South Regional Star Gaze, held deep in the dark pine forests of northern Mississippi. It was fairly early on the first night of the star party, but observers were already starting to drift away from the field. I was a little tired from the trip and the initial setup of equipment myself, but looking due south I noticed the bright beacon of Beta Ceti and remembered what lay in that part of the sky. I soon had my C8 SCT pointing at NGC 253—beauty itself in a wide-field Nagler eyepiece. Alone but for a friend or two, I just looked and looked and looked. At first my thoughts turned to boyhood views of this distant kingdom of stars, but before long I began to wonder what this galaxy must look like from the southern hemisphere. Thinking of southern climes suddenly led my tired but wondering mind to see this far-away star-swirl as a laden Spanish Galleon, bursting with gold, and sailing before the wind on a sea of endless night.

CHAPTER NINE

Winter

Tour 1
M78: Return of the Hunter

Walk outside on any early winter evening and your eyes and mind are filled with beauty. With the coming of this season our city skies seem transformed. The constant cold fronts passing though leave clear and, even in badly light-polluted areas, comparatively *dark* skies. Front passages sweep away particulate matter and moisture in the atmosphere, dramatically reducing light scatter. The sky appears darker because it actually *is* darker. Then there are the glorious stars of the winter Milky Way. Their names trip off our tongues as we stand awed: Castor, Pollux, Capella, Rigel, Betelgeuse. It's true that the summer sky actually features more bright stars, but the luminaries of winter, set against black-velvet skies and arranged in artful forms stand out better in our eyes and in our hearts.

Yes, there are many riches in the winter star fields for the urban observer, but there's one constellation amateur astronomers long for all summer, one constellation that stands apart even in the fiery winter skies, and one constellation that is the very essence of the winter heavens: glorious Orion. This legendary figure with his rectangle of bright stars, his blazing belt, and his mysterious sword is so distinctive that many people learn his shape even before that of the Big Dipper/Plough. For the telescope user, it's even better. Orion *is* the Milky Way. This great star-figure has those beautiful objects that draw us back to our home galaxy from nights of galaxy hunting: beautiful open clusters and nebulae of all kinds. There's also the added bonus that we have deliciously long winter nights in which to enjoy the Hunter's majesty.

The Milky Way's galactic star clusters can be beautiful and are easy to see in the city, but to me Orion has always meant *nebulae*. This constellation holds what are the Northern Hemisphere's best examples of the main classes of diffuse nebula: emission and reflection. Emission nebulae, glowing clouds of hydrogen gas, shine because they are "excited" by the radiation of hot nearby suns. Nebulae of this type are the birthplaces of the stars. As these clouds of hydrogen collapse under the influence of gravity and the shock waves produced by dying supernovae, massive newborns blaze into existence, blue–white "O" and "B" infant stars howling into the night with torrents of ultraviolet light. These high-energy ultraviolet photons turn a simple cloud of hydrogen into an indescribably beautiful artwork like the Lagoon Nebula glowing red in Sagittarius, or Orion's prize, the Great Nebula, M42, unsurpassed in the Northern Hemisphere for deep sky beauty.

Reflection nebulae are different. They are not as lovely as most emission nebulae. Or, at least, they're beautiful in a different, more subdued way. Reflection nebulae are lit, as the name implies, by the reflected light of nearby stars. They are not ex-cited as are emission nebulae, and appear in photographs as distinctly blue in color (emission nebulae tend to shades of pink and red). What's the reason for this differ-ence? It's due to the make up of these nebulae and the types of stars embedded in them. Reflection nebulae, like their emission cousins, are composed mainly of hy-drogen gas, but they also contain a large amount of dust. The real key, though, is the stars inside or near them. They are not luminous enough to excite the hydrogen into a glowing red/pink wonderland, but their light *is* strong enough to be well scat-tered by the nebula's dust. Why blue? They are blue for the same reason our daytime sky is blue: light of shorter wavelengths (blue) is scattered more readily under these conditions.

In photos, reflection nebulae are truly wondrous, shining with a dramatic icy-blue sheen. But, unfortunately for visual observers, especially those in light-polluted urban areas, they are usually terribly difficult objects. A prime example is the reflection nebulosity surrounding the star Merope in the Pleiades. Many visual astronomers go their entire lifetimes without being certain they've *really* seen the Merope Nebula. "Baby's breath on a mirror," doesn't begin to describe how elusive the Pleiades' nebula can be, even under the darkest skies.

Almost all reflection nebulae are at least as hard to see as the Merope cloud, but the winter sky does hold one of the few easy to observe examples of reflection nebula. One that is surprisingly impressive in city skies with small aperture telescopes: M78. That's our first stop this evening. Oh, don't worry—of *course* we'll visit the Great Nebula—but let's leave it for the end. We've got a lot of ground to cover first.

M78

M78 is almost as easy to locate as naked-eye-visible M42. You may not even need your trusty star atlas, though you'll probably want to refer to it if this is your first visit to this area of the Hunter. A glance at *Sky Atlas 2000*, your atlas of choice, or Figure 9.1, will reveal that M78 is about 2° 30′ north-northeast of brilliant magnitude 2.05 Alnitak,

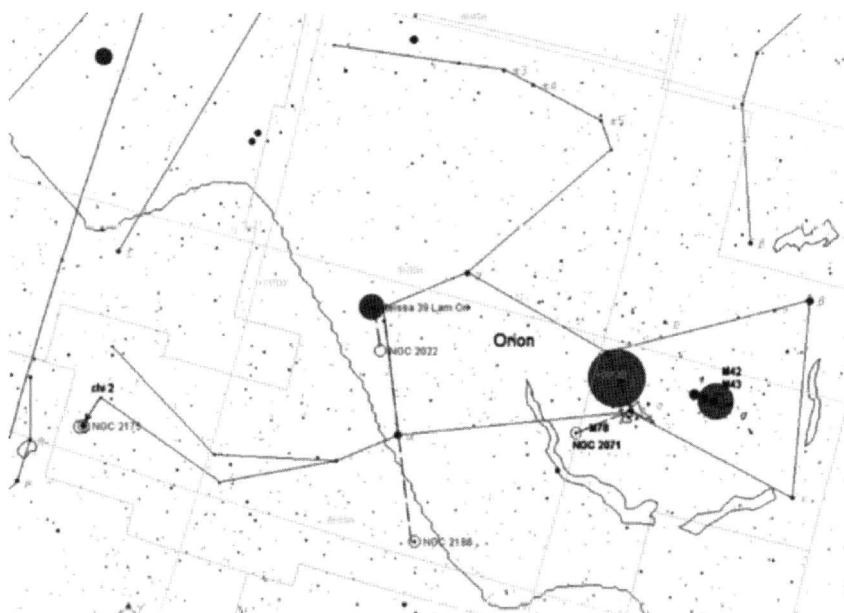

Figure 9.1. Orion, the Mighty Hunter.

Zeta Orionis, the easternmost star in Orion's belt. An easy way to locate the nebula without doing a lot of hunting is to visualize it as forming a 90° angle with the stars of the belt. While searching, use an eyepiece that yields approximately 50×. This will give a reasonably wide field of view, but should also provide enough contrast to make the nebula appear. When you've got the telescope positioned in the general area using your finder, switch to the main scope and look for a 10th magnitude double star in the field. These stars, 55″ apart, should look obviously enveloped in nebulosity, and resemble my drawing in Figure 9.2. You may have to use averted vision to see the nebulosity easily.

If you have trouble locating the right spot, go back to your chart to make sure you are in the right area. I've often found that when I cannot locate an object after an extensive amount of sweeping, I'm usually many degrees away, not even in the right part of the constellation, not just a field or two off. If you think you do have the double star in your field, but can't make out the nebula, switch to an eyepiece that gives you about 100 150× to help spread out the background sky glow.

The true size of M78 as revealed in photographs (see Plate 50) is 8′× 4′, but you will undoubtedly see less. You should, however, be able to discern over 1′ of nebula with an integrated magnitude of about 8.0 in the form of a circular patch around the double star. I was frankly amazed at how good this object looked in the 80-mm *f*/5 refractor on an above-average night in the city. It was not at all difficult to see, and the nebulosity was prominent enough to almost assume the character of a "deep sky showpiece." My observing logbook notes:

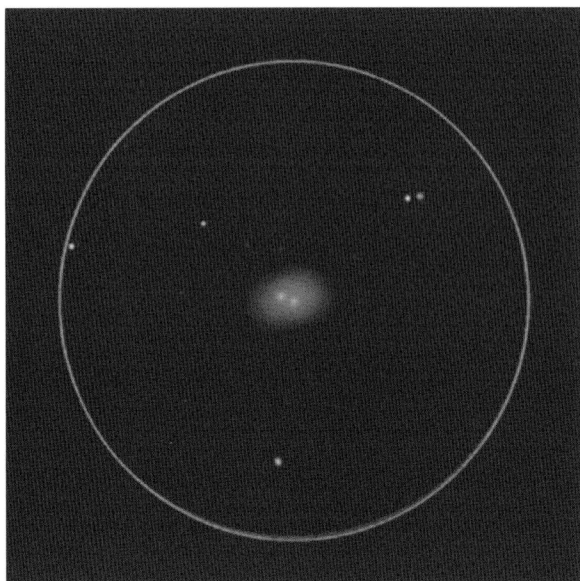

Figure 9.2. M78 is a little fuzzy cotton ball in amateur scopes.

Once found, the nebula is very easy to make out in the 80-mm refractor. Certainly easier than M1 from these skies. Amorphous and elongated. With some difficulty, using averted vision, it seems to morph into two lobes of nebulosity rather than one round patch, at times appearing peanut shaped and looking almost like the Little Dumbbell Nebula in Perseus.

Like many dimmer nebulae, you may need to wait for an especially good night and for Orion to be on the meridian for much of a look at M78. While the Short Tube 80 has revealed the nebula nicely on better than average evenings, there've been plenty of times when it's been invisible in it or in the slightly larger 4.25-inch Newtonian. It has never eluded me with an 8-inch scope, however.

I've observed M78 on countless evenings with a wide range of telescopes and, until recently, if you'd asked me, I'd have told you that I have seen just about everything this little cloud has to offer and know exactly what to expect from it in everything from a 60-mm refractor to a 30-inch Dobsonian. Naturally, it's when you start entertaining grandiose ideas like that that the sky decides to teach you a lesson. Out with my 11-inch SCT on an only fair night, I recorded in my log that

The main patch of nebulosity around the pair of stars is visible, but subtle. Not easy. Requires averted vision in the 22-mm Panoptic eyepiece. A 12-mm Nagler makes it even less obvious tonight.

In other words, the nebula was singularly unimpressive. Normally, I would simply have moved on to something else. Instead, I screwed a UHC narrowband light-pollution filter onto the eyepiece. Why did I do that? Don't ask me. As noted in the accessories section, light-pollution filters *shouldn't* do a darned thing for a reflection nebula. They dim the light of stars, and since reflection nebula "shine" because of reflected starlight, they should be made invisible by a UHC filter. Nevertheless:

> M78 is much improved by a UHC light-pollution filter on the 22-mm eyepiece! It is significantly more visible with the filter than without. Without the filter, it is a round or slightly elongated glow; with the UHC it is an irregular, lobed, and relatively large structure.

Why? As is the case with many reflection nebulae, there is at least a minor emission component to the nebula; M78 is apparently a mixture of emission and reflection nebulosity. There is enough emission nebulosity involved in M78 to make an LPR filter like the UHC fairly effective. The main point of this story? That even after decades of observing, you shouldn't become too complacent about what you can or cannot do with a telescope or how you should use one. If somebody tells you that you can't use your telescope "that way," go right ahead and try. You may be surprised at the results.

Really *study* M78. Just don't find it, glance at it, and move on to your next challenge. Spend some time with this object. Make a drawing. Push yourself to see just that single additional bit of detail. I guarantee, the more you look, the more you'll see. If the main nebula is easy, can you make out an additional object in this field?

NGC 2071

NGC 2071 is a smaller, "detached" portion of the M78 nebula 15′ NNE of the main cloud and visible in the same medium-power field as the main nebula. It's dimmer than the main patch, and only about half its size at 4′ in diameter. Like M78, it surrounds two stars, both, like M78's stars, of 10th magnitude. In larger apertures, in fact, NGC 2071 looks just like M78 does in a 6-inch telescope. I've seen NGC 2071 in 10-inch and larger instruments on *very* good nights in the city, but it is challenging. Very challenging. Which doesn't mean *you* might not be able to bring it back with a 6-inch or 8-inch telescope.

M78 was discovered by Charles Messier's contemporary, Pierre Mechain, in 1780. Despite its somewhat pedestrian appearance in our telescopes, in reality it is a huge and wondrous thing. M78 is believed to be about 1500–1600 light years away and is thus about 2–3 light years across. In other words, this "little" DSO is much, much larger than the main part of our Solar System. While our eyes usually only see one or two dim comet-like objects here, long exposure photographs reveal a much more extensive area of nebulosity as seen in Plate 50.

Color images are especially interesting because they show subtle tinges of pinkish red in addition to M78's overall reflection-nebula blue. That is because there is an obvious emission component here, as I discovered when I attached my UHC filter to the eyepiece. There are enough energetic photons from the stars buried in M78 to encourage at least some atoms to fluoresce. This combination is seen in much more dramatic form in Summer's Trifid nebula, which is so bright that it's often possible to see subtle hints of color difference in it's reflection and emission parts visually with the aid of a medium-large scope.

This whole area is filled with subtle patches of nebulosity. Most, like the recently discovered or rediscovered spot, "McNeil's Nebula," found by U.S. amateur Jay McNeil on one of his CCD images of M78, are very difficult. McNeil's nebula has been referred to as a "visual object," but only from dark skies with sizeable aperture.

After half an hour or more of searching for and observing M78 and its dimmer kin, I'm ready for a break. Time for a sip of coffee, a stretch, and a glance at my star charts to reacquaint myself with the next stop on our night-journey.

Where to next? Given Orion's spectacular appearance and location alongside the Winter Milky Way, you'd think that it would just be brimming with luscious deep sky delights. It is, but only if you're out in the dark countryside with a very large telescope. Open one of those pretty, full-color coffee-table astronomy books and Orion's tantalizing and legendary diffuse nebulae spill off the pages: the Flame, the Horsehead, the Running Man. Beautiful images that fire my imagination and my appreciation for the deep sky. Unfortunately, as stimulating as they are in pictures, these nebulae are dim and subdued visually even with large apertures from the darkest sites. In the city, they are just plain *not there.*

What about star clusters? Orion has scads of these, but most are dim and unimpressive in medium-sized telescopes in light-polluted skies. There are no Messiers among them, and Orion's star nests pale beside those of nearby Gemini and Auriga. Planetary nebulae? One or two. But nothing as good as the Ring. Not hardly. Or even as good as M97, Ursa Major's dim Owl. Galaxies? Surprisingly, given Orion's Milky Way location, yes, but none that are anything more than the dimmest of smudges under gray skies. Is there *anything* else "good" to see other than the Great Nebula and M78? From town? Not as much as you'd expect, no, but there are a few detours the diligent observer can make on the way to M42 and the wonderland of the sword. All these objects are identified in *Sky Atlas 2000.*

NGC 2186

NGC 2186 is a fine, small open cluster 4′ across, shining weakly at magnitude 8.7. In the 11-inch Schmidt Cassegrain, I saw as many as 20 tiny stars arranged in a crescent shape. This group probably requires an 8-inch telescope to see well, though it's no doubt detectable in much smaller instruments as a small hazy patch. Look for it northeast of Betelgeuse. Start at Phi Orionis, 4° 35′ west of Betelgeuse, draw a line from Phi and through Betelgeuse and on into space for the same distance, 4° 35′, and you'll find NGC 2186 nestled in a medium rich star field.

NGC 2174, NGC 2175, and NGC 2022

NGC 2175, another open cluster, is large and sparse—I identified 10 stars that appeared to be cluster members scattered across 20′ of sky. The real attraction here isn't these dim pinpoints, though, it's the large emission nebula, NGC 2174, located to the southwest of the cluster. This nebula *may* be just barely visible in urban skies with a 10-inch telescope equipped with a UHC filter on an *excellent* night. It is much easier to see in the city with a 12-inch telescope, but is not a routine catch. For that matter, it's not overly easy even with large scopes, 15 inches and above.

In the C11, I was able to detect a *very* faint haze extending outward from a magnitude 7.5 star. The nebula, like the star cluster, seemed to be about 20′ in size. The cluster is easy to locate in almost any scope, as it is comparatively bright at magnitude 6.8 and lies only 1° 22′ northeast of the bright star Chi2 Orionis. As for the nebula, you're on your own. Use a wide-field eyepiece equipped with a narrowband nebula filter, scan the area to the south and west of the cluster, and keep coming back to this spot on superior nights.

NGC 2022, a small planetary nebula, is located 2° east of magnitude 3.54 Lambda Orionis, an attractive double star and one of the three suns that form the small triangular asterism representing Orion's "head." I'd never observed this planetary until recently because its small size (19″) and its dim magnitude (12.4) didn't seemed to make it worth bothering with—even from dark sites. When I finally got around to having a look at this nebula, I was surprised by how nice it looked. It *is* small, but even at 100× it was easily identifiable as a planetary nebula. It also seemed much brighter in an 8-inch telescope than the magnitude figure would indicate. At 200×, it was rather impressive in the 11-inch SCT, appearing as a small gray ball. Some detail in its tiny disk seemed just on the edge of visibility. I didn't see a central star, however, and it was really just a featureless fuzzy. How much scope do you need for it in the city? This planetary was challenging in an 8-inch scope, but I have no doubt that it *could* be doable in a 6-inch scope. I tried an OIII filter, but noted little improvement in visibility or detail.

M42

Are you ready? You won't need a chart to find the evening's final object. M42, the Great Orion Nebula, is easily located by anyone who can see Orion's Sword. The middle, "fuzzy" star is the Great Nebula blazing across the light years, penetrating even the worst light pollution with ridiculous ease. I won't bore you with a long discussion M42, but I will make a few suggestions that may enhance your pleasure in viewing this marvel.

OK, so you've seen this object a million times before. Why not just take a quick look at M42 and move on to other things? Everybody with a telescope knows just what it looks like, right? Not necessarily. How long has it been since you've made a *detailed* examination of this huge cloud? Do you at least *try* to make drawings of this object (see my impression of this monster nebula in Figure 9.3)? Do you view it with a wide range of magnifications? Every time I look—*really* look—at M42 I seem to find something new. It looks great in tiny telescopes, and bigger instruments only enhance my feeling of wonder. Don't neglect the rest of the Sword, either. While you probably won't see the fascinating complex of reflection nebulosity, NGC 1973/1975/1977, the Running Man Nebula, there are many pretty stars and clusters scattered all up and down the Sword.

If there's a deep sky object (DSO) that's as impressive in the city as it is in the country, it's M42. Sure, you lose some of the outlying wisps of nebulosity in urban skies, but a light-pollution filter can bring much of that back. It's hard to suggest the best telescope to use on this marvel. It looks just as good, frankly, in a Short Tube 80-mm refractor as it does in a big Schmidt Cassegrain or Dobsonian. The big scopes

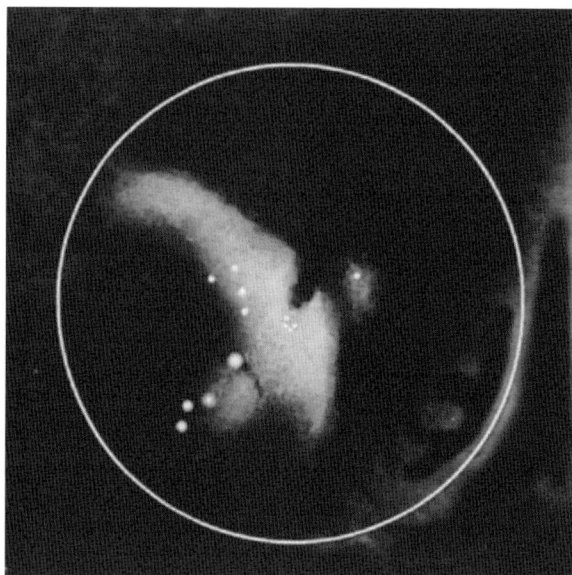

Figure 9.3. The Great Nebula.

allow me to zoom in on the details of the nebula's interior, but the small one allows me to see the whole cloud set in the wonderful sword. Aside from a light-pollution filter, the one thing that will add more to your joy in viewing M42 than anything else is a wide-field eyepiece. This is one time when the City Lights astronomer wants low magnification. Even in the city, the nebula stretches all the way across the field of a 35-mm Panoptic eyepiece and beyond in my C11.

In addition to the dark intrusion in the nebula's Northern side, the "Fish Mouth," and the bright multiple star system, the Trapezium, Theta 2 Orionis, look for the "wings," the streamers of nebulosity that flow east and west. On the northern edge of the nebula, surrounding a magnitude 7 star, is the small, detached cloud, M43. In the city a 10–12-inch scopes reveals that M43 is comma-shaped and, on the very best evenings, hints that it's crossed by dark dust lanes. However big or small a telescope you own, turn it to M42 frequently. Don't be embarrassed to keep coming back to this easy object at every opportunity. It's a mind-blowing sight, always.

Tonight's Double Star: Meissa, Lambda Orionis

Lambda, one of my favorite double stars, is challenging without being trying, and is one of the first double stars I show my university astronomy students. It looks like a double "should": two close yet distinct points of light. The primary is a bright magnitude 3.6, and lying 4.4″ away is a dimmer secondary star shining at magnitude 5.5. The primary is blue–white, and the secondary is just plain white, though some observers report a trace of "olive green" in it, an illusion caused by the presence of the much brighter main star. Though usually split by a 6-inch telescope, poor conditions can make larger aperture necessary for success with Meissa. Lambda is easy to locate, being, as mentioned earlier, one of the three stars in Orion's triangular head, just to the northwest of Betelgeuse and Bellatrix.

> Don't ever discount the value of your most wonderful optical instrument—your unaided eyes. I think one of my most memorable views of Orion's blazing suns was with my unhindered eyes. One unseasonably warm night in the late 1960s or early 1970s, I sat in a country field with a thoughtful young woman and drank-in the majesty of the Celestial Hunter. Ever after, Orion has been associated in my mind with the scent of patchouli and the sound of folk guitar, music favored by a generation now growing old, but forever young in the contemplation of the distant stars.

Tour 2
Challenges for Deep Winter Nights

What shall we look at tonight? As always, a *variety* of deep sky denizens if that's possible. After observing for a few years, most deep sky fans tend to focus on a single type of object, often to the exclusion of all else. No self-respecting galaxy hunter, for example, would be caught dead wasting his/her time on mere open star clusters. But this narrow concentration means you miss some wondrous views. I'm the original astronomy dilettante; I'll look at *anything*. In part, it's the contrast between the very different animals in the deep sky zoo that makes my nights so interesting.

With this in mind, I decided to select several radically different destinations for tonight's expedition. Perseus is a natural hunting ground if you're after contrasts. This is one constellation that has it all: numerous star clusters, elusive diffuse nebulae, intriguing planetary nebulae, beautiful and mysterious double and variable stars, and even a challenging galaxy or two. I eventually selected three very diverse and lovely stops for tonight's itinerary: open cluster M34, planetary nebula M76, and galaxy NGC 1023.

M76

Our first destination, M76, the Little Dumbbell Nebula in Perseus, is not overly hard to find, though it is, at magnitude 11, one of the dimmest of the Messiers. Its $3.0' \times 2.0'$ size makes its surface brightness fairly high, but I won't call it "easy." It's time for *challenges* after spending most of the winter in the house or stumbling across easy open clusters. Figure 9.4 should lead you to the Little Dumbbell without too much hassle, however. Look for the nebula slightly less than a degree northwest of Phi Persei. This star is not very bright at magnitude 4.07, but it should be easy enough to land on if you hop there from bright Gamma Andromedae (magnitude 2.36) and 51 Persei, which shines at magnitude 3.57, and which is only $2° \ 15'$ from Phi.

Center Phi Persei in a low-power, wide-field eyepiece and slowly sweep one eyepiece field's distance toward Delta Cassiopeiae—very slowly. You'll be looking for a small puffball. There is a 6th magnitude star $12'$ east-southeast of the nebula, the only even marginally prominent star in the immediate area. If you have difficulty finding the M76, stop for a moment, get your bearings, take a deep breath, and try again. Remember to examine each field carefully since this planetary may be quite hard to see in a small scope under really poor conditions.

When M76 finally appears in your eyepiece, you may or may not be able to make out its dumbbell shape. In small scopes, you'll probably find it appearing as a slightly elongated or rectangular patch rather than as two distinct lobes. Even on the darker nights when M76 shows off some detail, don't look for it to resemble the big Dumbbell, M27, in Vulpecula. M27 looks more like an apple core than a dumbbell, while M76

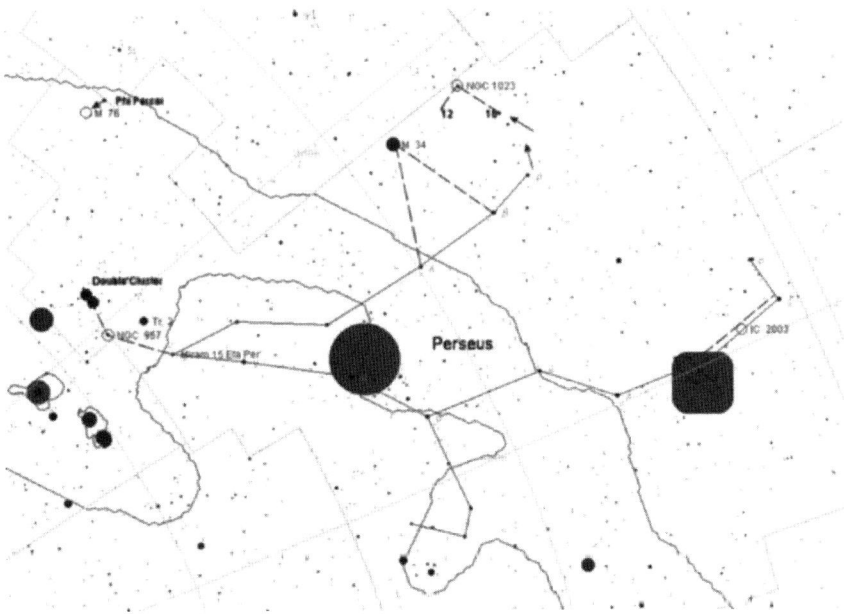

Figure 9.4. The star fields of Perseus.

is composed of two round lobes that are in contact, actually making it look more like a piece of exercise gear (in a medium-sized scope) than the big Dumbbell does. How hard is this double-lobed shape to see in the city? Not terribly hard in 8-inch and larger instruments equipped with OIII filters—which help considerably with this planetary. The challenge at 12 inches of aperture and above is not to see the dumbbell shape, but to be able to detect that one lobe is brighter than the other and, beyond that, to see vague hints of the outlying haze of faint nebulosity that surrounds this object.

Viewing with a very small telescope? Don't automatically assume M76 will be invisible. I have been amazed at how easy this dim planetary nebula can be on the right night. On a good post-cold-front evening, I had no problem observing it with my OIII equipped 60-mm ETX 60:

> I was surprised and pleased to be able to see the Little Dumbbell from in-town with 60-mm of telescope aperture. Required the OIII filter to see well. Removing the filter, I found I could still see the nebula, but just barely. Could not resolve it into two lobes, but it was otherwise easy in this scope, appearing as a uniform and elongated oval of nebulosity at 75x.

On poorer nights, it could be difficult even with the larger aperture of the 4.25″ Newtonian:

> Interesting, but really just at the limit of visibility in this aperture under these skies. Diffuse and round looking.

The long focal length and medium aperture of my Nexstar 11 SCT proved to be a good combination for M76:

> The Little Dumbbell. Exquisite in a 26-mm Plossl, even without a filter. Lovely field. Both lobes easy at low power. Some suggestion of faint streamers of nebulosity wrapping around one of these lobes.

Even on frustrating nights, I found M76 to be more than worthwhile. Especially if I stuck with it rather than quickly moving on. Extended observation with higher powers would almost always reveal at least a hint of it in my smallest scopes under the worst skies (including high humidity and incipient fog). It has occasionally been invisible in my Short Tube 80 and ETX 60, but I have never failed to bag it with a 6-inch scope with the OIII filter installed.

M76 is a Vorontsov–Velyaminov type 3(6) planetary nebula (irregular disk/ anomalous form), which is located about 1700–3000 light years from Earth. Like almost all planetary nebulae, M76's distance is not well known. Photographs taken of this object by very large telescopes show that in addition to its curious non-disc structure, M76 is wreathed in many faint streamers and tendrils of nebulosity. These streamers are rather difficult to see visually, but are readily apparent in CCD exposures with amateur telescopes like the 8-inch SCT that took the photo in Plate 51.

M34

The little Dumbbell *was* hard to find in sodium-pink skies tonight. But the next stop, M34, is much easier. After M76 you should almost be able to find M34 with your eyes shut. Using a detailed chart, search for this object 5° west-northwest of the famous eclipsing variable star Algol, Beta Persei. This magnitude 5.2 open cluster should be at least barely visible in your finder even in pretty heavy sky glow. Under really dark skies, M34 is obvious to the naked eye. At 35′, it just fills the field of a low-power eyepiece, so stick with your longest focal length ocular for the most pleasing look at this sprawling star-nest. You may want to try higher power to see if you can pick up some of the cluster's less prominent stars, but be forewarned that they are at magnitude 13 and dimmer.

I enjoyed viewing M34, though I must agree with the late Walter Scott (Scotty) Houston who thought the cluster was "sparse." Looking through my log, I see that on the same night I viewed M76 with the ETX, I also took a quick look at M34.

> Easily seen, fills a half degree field. M34 isn't super-spectacular, but is easily seen, found, and is rather attractive. Concentrated toward the center, basically oval in shape, though I sometimes have an impression of an almost "spiral" shape being formed by the brighter stars. No nebulosity seen. About 40 stars seen at 45×.

Under the Trumpler classification scheme for open clusters, M34 is class II 3 m (detached with weak concentration toward center, large range in brightness, moderately rich). At a distance of 1500 light years, it is approximately 20 light years in extent. This group's age of 100 million years or so makes it somewhat elderly as open clusters go, being senior to both the nearby Double Cluster and the Pleiades.

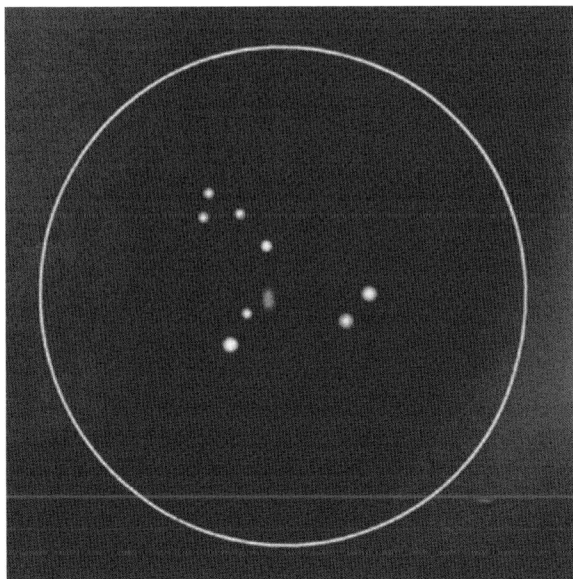

Figure 9.5. NGC 1023, the Little Spindle.

NGC 1023

Our final featured deep sky tour stop for this evening, galaxy NGC 1023, the Spindle Galaxy, seems out of place here in the midst of the winter Milky Way, but some galaxies do manage to make their presence known not far from the "zone of avoidance" created by our galaxy's huge, dust-swollen body. NGC 1023 isn't overly difficult to see most of the time, though you should try to wait for a good night if your scope is small or your skies are very bad. It is, in fact, rather spectacular as small NGC galaxies go. I recall that the first time I hunted down this object from light-polluted skies I surprised myself by finding this 11th magnitude marvel in less than 5 minutes. Maybe because the bright stars 12 and 16 Persei (magnitudes 4.9 and 4.2, respectively) form a little triangle with NGC 1023, making it trivial to pin down.

To find this galaxy, star hop your way up from Algol's neighbor, Rho Persei. Hop from Rho, to 20 to 16 and stop. Looking at your star atlas, observe that NGC 1023 is 2° 7′ to the west from 16 Persei, and forms a triangle with it and 12 Persei. 20, 16, and 12 are all in the magnitude 4–5 range, and should be easy for your finder. The first time I hunted NGC 1023, I did have to pay strict attention to what I was doing since, under my heavily light-polluted skies, all I could really see of the galaxy was its small, elongated core, which is visible in my 4.25-inch scope drawing in Figure 9.5. Even this "bright" central area was best seen with averted vision.

On a decidedly average night of seeing and transparency, my notes recall that NGC 1023 was:

Visible in the 4.25-inch scope, with averted vision. Seems elongated E/W at 90×. No sign of a stellar core or outer envelope. It disappears completely when I look directly at it, and higher powers don't seem to help much at all.

In scopes of 8 inches of aperture and above, NGC 1023 can be distinctly variable. One night it'll look "ho-hum," as above. At other times it will begin to give up some detail to the methodical observer. In particular, the elongated saucer shape of its main body, which is responsible for its "spindle" nickname, becomes obvious.

In the C8 at f/10 with the 12-mm Nagler and TeleVue 2× Big Barlow, NGC 1023 is fairly obvious and surprisingly large at 300×. In addition to the bright, oval core, a fairly large and strongly elongated expanse of nebulosity is visible tonight along with a bright, sub-stellar nucleus.

NGC 1023 (Plate 52), which astronomers have classified variously as a Hubble E7p galaxy or an S0, appears to be a lenticular galaxy with a disturbed-appearing nucleus—it is included in Halton Arp's catalog of "peculiar" galaxies. This part of the constellation contains quite a few dim galaxies, a number of which lie just to the east of NGC 1023, and most of which are beyond the range of any but the largest urban scopes.

Stimulating Side Trips

NGC 957

NGC 957 is a better-than-average open cluster that's easy to find 1.5° east of NGC 884, the easternmost of the famous Double Cluster "twins." To find it, scan east and just a little north of the Double Cluster, and be on the lookout for a fairly conspicuous triangle of 7th magnitude stars. You may want to use a detailed, small-scale chart to track it down, as this is a rich area, and there are many "cluster-like" clumps of stars everywhere that can deceive you. While NGC 957 is a *good* cluster, it's big, a sprawling 11.0′ in extent, and it may not be immediately obvious against a rich background. At an integrated magnitude of 7.6, the 20 or so stars you see here, arranged in a shapeless pattern elongated east–west, are not a challenge once you're in the right spot and know what you're looking for.

IC 2003

This is far from the easiest DSO we've run down tonight, but this tour *is* about challenging objects. This planetary nebula is faint at magnitude 12, but that's not the problem. The trouble is that it's *small*, with various sources giving it a diameter of from 4″ to 8″. Due to this minute size, it's bright, at least, and is certainly more prominent than NGC 1023. It's not a challenge for your finding skills, either, being located almost exactly equidistant between the bright stars Xi and Zeta Persei. It is situated in the same area as the well-known—if notoriously dim—California Nebula. The problem is distinguishing it from an anonymous field star. How do you pick it out? As with other small planetaries, two things are critical, high power, and an OIII filter. I found I could detect it—barely—as a slightly fuzzy star at 166× in the 8-inch *f*/5 Newtonian, but required an OIII filter to be sure. The OIII dimmed the other field stars while making the nebula *slightly* more prominent. Identification can be easier at higher magnification on nights when the air is still and you can pour on the power. The nebula became quite obviously nonstellar at 350× and above in the C11 SCT.

Tonight's Double Star: Miram, Eta Persei

Miram is a very good-looking pair of stars color-wise, with its supergiant primary star being an intense orange-red, and the secondary a cool blue. The blue companion star is much dimmer at magnitude 8.5 than the magnitude 3.8 primary, but their wide separation, 28.3″, means resolving them, observing them as two separate stars, is not as difficult as you'd think, and quite possible in small apertures if the seeing is steady. Eta is rather trivial to locate since it's a part of Perseus' stick figure constellation pattern. It forms a triangle with Tau and Gamma Persei. These three stars are often used as a "pointer" when searching for the Double Cluster.

> Feeling a bit of a strain from locating and viewing yet another faint fuzzy, I pull my eye away from the eyepiece and for a moment just look up at the skies. Overhead, the distant, alien stars of the Perseus OB-3 association blaze away. After a while, my gaze settles on the spot where I know galaxy NGC 1023 is located. Though I can't hope to see this dim and distant creature with my unaided eyes, I can see it with my mind's eye, and suddenly, with a rush of something that almost feels like vertigo, I sense the true three-dimensional reality of the cosmos. Perseus' bright stars are my close neighbors, nearby friends, no more alien than my Earthly acquaintances when compared to the really distant and peculiar NGC 1023.
>
> We are all of us—the residents of the Milky Way—huddled together for comfort against this immense darkness. Even NGC 1023 is amazingly nearby when we consider the inhuman distance that lies between us and the quasars. One of the most rewarding tools of the amateur astronomer is imagination. Our little telescopes give a sometimes dim outline of the truth of things, but our minds can fill in the details of the realities of our universe.

Tour 3

A Surprise Planetary

"The Moon will be rising soon and it's cold out here. How about something a little easier than M76 and NGC 1023 for this outing?" I *was* thinking we should catch Pisces' M74, which is beginning its descent into the west, but that dim, face-on Sc spiral, while doable from urban areas—I've seen it with the 4.25-inch scope with major effort—is tortuously difficult. More than once, even on the best nights for City Lights astronomers, I've come away from my telescope scratching my head and wondering whether I've *really* seen M74 or not. OK, I'll give you a break. Here's a set of easy, quick prizes, a trio of beautiful Messier objects, which, due to their southerly location in an out-of-the-way constellation, few casual observers ever visit. And one holds a delicious surprise!

M46

Although M46 resides within the borders of the somewhat obscure southern star pattern, Puppis, the Poop Deck of the obsolete giant constellation *Argo Navis*, the ship of Jason, it is easy to find. Don't use the horizon-dimmed dim stars of Puppis as your guides, though. Puppis is populated by numerous bright stars, including four brighter than magnitude 3, but for most Northern Hemisphere observers they don't look very bright at all due to their low altitudes. Instead, find your way by the brilliant beacon of the Dog Star.

This open cluster resides 13° 45' east of Sirius, Alpha Canis Majoris. The only complication is the possibility of mistaking M47 for M46, M47 is *another* nice open cluster that lies only 1° 30' to the west of M46. Look at Figure 9.6, and you'll see that M46 (and M47) form a near equilateral triangle with bright Sirius and mag 2.2 Rho Puppis, one of the more distinct of Puppis' stars. Minimal optical aid, finder or binoculars, should easily show both M46 and M47. Just remember that M46 is on the east and M47 is on the west and take into account the probably inverted image of your finder, and you can't fail.

Once you have the cluster centered in an eyepiece, you'll be pleased at how pretty it is. Consisting of at least 150 stars between magnitudes 10 and 13, this manageably small beauty is less than 30' in extent and puts on a nice show in any telescope. Imagine, though, how wonderful a sight it is under dark skies in the Southern Hemisphere, since it is situated right in the heart of the marvelous Puppis Milky Way. Even on a very poor night in a Northern Hemisphere city, however, I recorded in my log entry (and my drawing in Figure 9.7) for the 4.25-inch Newtonian that M46 was:

> Amazingly lovely and compact. Scattered clouds and haziness tonight, so this cluster is a little dim. Outstanding nevertheless, with at least 30 stars visible in the field of a 25-mm Kellner eyepiece at 48×. The basic shape of the cluster is round or slightly elliptical, and I can detect a faint background glow of unresolved stars.

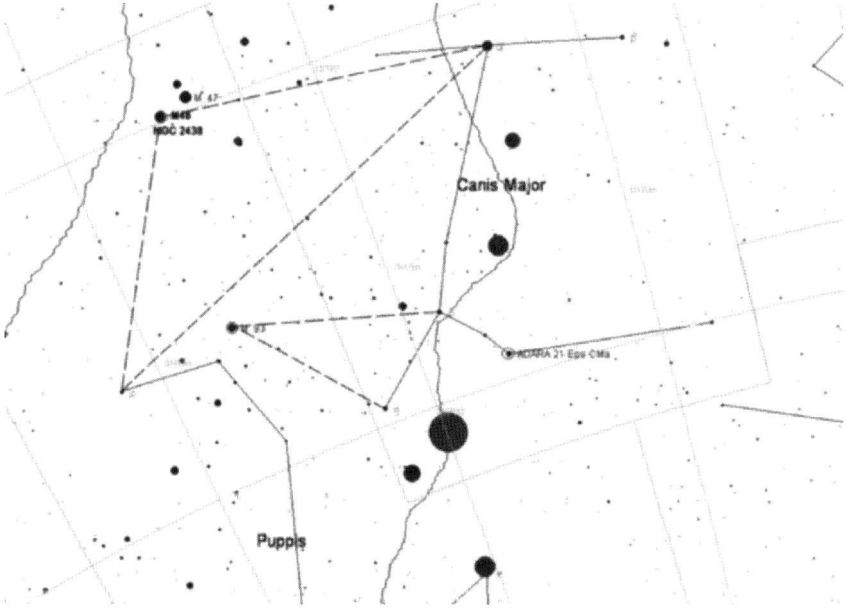

Figure 9.6. Puppis and the Big Dog.

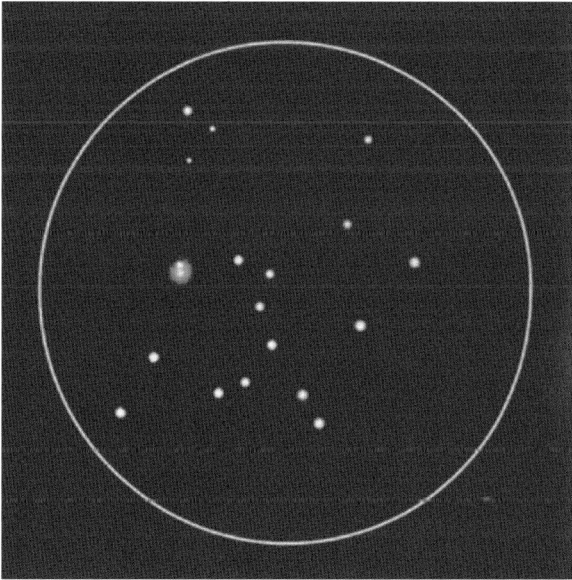

Figure 9.7. M46 and its hidden prize.

Open clusters are nice, but there is a certain undeniable ho-hum factor to *another* half-degree group of semi-dim stars. However, a patch of bright stars is not the *only* attraction here. Once you've formed a general impression of this cluster, look closer. About 7′ north of the center of this cloud of stars, you'll find M46's "surprise," the planetary nebula NGC 2438

NGC 2438

The magnitude of this small planetary, 11.0, sounds forbidding, but its small size, 1′×1′, makes it nice and bright. The problem is not its magnitude, really, but as when we searched for IC 2003, its small size. Once again, apply enough magnification so that it's distinguishable from cluster stars without dimming it so much that it becomes invisible in your scope from your skies. If it isn't immediately obvious, search the field carefully for a "star" that seems a *little* fuzzy with averted vision. If you still can't see it, increase your magnification further. While this makes the view of the cluster much less attractive, magnifications above 150× begin to make the nebula more obviously nonstellar. If NGC 2438 is *still* not seen, check to make sure you're not *really* looking at *M47* rather than M46, a mistake I've made a time or two. If you have an OIII filter available, use it, as this nebula responds well to one.

I would classify NGC 2438 as a "moderately difficult" object from the city with a 4-inch or smaller telescope (it's barely visible as a small blob in my drawing), but it's definitely possible on the right night, so keep after it. In this aperture range, you'll see a tiny gray disk superimposed on a pair of close, bright cluster stars. Averting your vision makes the nebula easier to see, and you'll note that even this small object displays the "blinking" effect typical of planetary nebulae in smaller telescopes.

The key to enjoying this nebula is going after it on the right night. On poor evenings, NGC 2438 can be surprisingly hard to detect in an 8-inch scope. On all but the worst nights, however, this planetary is easy in 10–12-inch telescopes and sometimes even lets go of a few details:

> In light pollution and fairly heavy haze, M46 is easy, but, even in the C11 equipped with an OIII filter, the planetary, NGC 2438 is not immediately obvious. I am finally able to pick it up by waiting for Puppis to rise a little higher in the sky. As the evening wears on, the haze seems to clear off a little, making things even easier. I still can't seem to make out the nebula's disk easily with the 22-mm Panoptic, though. Going to the 12-mm Nagler at 220× makes the nebula stand out like a slightly greenish traffic light amid the yellow headlights of the cluster stars. At this magnification (again with the OIII filter), NGC 2438 is a fairly large gray–green disk abutting a pair of bright stars. Averted vision reveals a ring shape, like a miniature M57, with a surprisingly dark center.

M46, which is shown pretty well in the Digital Sky Survey image in Plate 53, is located 5,000 light years away, or so we're told, and is therefore 30 light years across. When researching this object, I was a little disappointed to find that NGC 2438 is most *probably* a foreground object, being, perhaps, 3,000–3,500 light years from dear, old Earth, and not actually involved in the star cluster.

M47

While you're in the neighborhood of M46, you might as well hot-foot it over to M47, just 1° 18′ away, maybe just on the western edge of your eyepiece field if your scope can deliver a degree of field with a low-power eyepiece. M47 is not quite the cluster that M46 is, I'm afraid. It's sparser and doesn't contain any added attractions like M46's planetary. But it's a nice enough DSO in its own right and deserves a few minutes of your telescope time. This 29′ diameter magnitude 4.3 cluster is easy to distinguish from the background even in the busy Puppis star fields.

When you've got M47 in the main scope's eyepiece, crank *down* the power as much as you can. Even for me at a latitude of 30° north, M47 and Puppis are always near to the horizon and in some pretty fierce sodium-arc light pollution and haze, so there's a limit to how low I can go before the brightening eyepiece field at lower power devours a good portion of this cluster's stars. The 35-mm Panoptic was guilty of this, with a 22-mm Pan or a 26-mm Plossl providing a decidedly better look at M47:

> An attractive enough field in the 22-mm Panoptic and Nexstar 11. M47 appears sparser by far than M46. The cluster is defined by about 10 attractively bright stars. Maybe about that many or a few more dimmer stars are also seen. One nice double is involved in the cluster. M47 is shapeless and scattered, but still makes a nice stop-over in this rich area.

M47 was discovered by Charles Messier in February of 1771. While there's no doubt that he independently discovered it, there is some evidence that it may have been recorded as much as a century earlier by other observers. Unfortunately, Messier wrote M47's coordinates down incorrectly, so this was a "lost" M object until NGC 2422, the cluster's original catalog designation, was convincingly identified with Messier's M47 in the 1930s. M47 is assumed to be 1600 light years away, and about 10–12 light years in size.

M93

Embedded as Puppis is in the Milky Way, it shouldn't be surprising that there's another outstanding galactic star cluster to be found in this constellation, M93, which is considerably more attractive than either M46 or M47. No, it doesn't harbor a planetary nebula, but it does possess an attractive compact core that makes it another open cluster that's reminiscent of Sagitta's loose globular, M71. In fact, I keep running across open clusters that look like very loose globulars so often that I've concluded M71 is not nearly as unique as I used to think. There is no question about M93's true object type; it is a young cluster, "only" about 100 million years old a positive infant compared to even the youngest globulars.

With a magnitude of 6.5, M93 is dimmer than either of its companion clusters in Puppis, but it's also smaller at 22′, and is quite impressive though it's considerably farther south than either M46 or M47 at a –23° 51′ declination. To locate it, you can hop from Xi Puppis. This magnitude 3.34 star *may* be fairly distinct in your skies, and if so, finding M93 is easy. The cluster is a mere degree and a half to the southwest. If Xi is invisible from your site, use the bright feet of the dog, magnitude 2.45 Eta Canis

Majoris and magnitude 1.84 Delta Canis Majoris. M93 forms a near 90° triangle with Eta and Delta.

In the eyepiece, M93 is remarkable, yielding as many as 50 "easy" stars to a city bound 4-inch telescope at high power. The most attractive aspect of the cluster, without doubt, is its small, compact core:

> In the 4.25-inch f/10 reflector this is a very good cluster at 48× with the 25-mm Kellner. Remarkable, in fact, despite being way down in the thick air to the south. There's a triangular core composed of a sprinkling of many, many tiny stars that show up well despite my observing site, which is located near the site of a large shopping mall. Many medium-bright to dim outlying stars visible, too, surrounding the cluster's distinct arrowhead-shaped center.

M93 is one of Charles Messier's personal discoveries, having been observed and catalogued by him in March of 1781. This cluster's Trumpler classification is normally given as I, 3, r, making it "detached, strong concentration toward center; large range in star brightness, rich."

The Other Side of the Sky

If you've waited long enough for Puppis to rise, or if it's late enough in the Winter for the Poop Deck to be over the horizon at a reasonable hour, a look over your shoulder to the South and West will reveal that Orion and company are skittering surprisingly far into the West. Orion is still well placed for observing, but why not trip on over to the obscure if colorfully named constellation, Monoceros, instead? The Unicorn resides just to Orion's east.

NGC 2237

"Colorfully named" is one way to describe the Unicorn. "Subdued to the point of near invisibility in the city" is another. Monoceros is composed of a vague "W" pattern of dim stars scattered across a fairly large patch of sky—481.6 square degrees lie within the borders of this constellation. Within these precincts, hidden among the dim suns, you'll find some marvelous clusters and nebulae. In fact, it's nebulae that most observers associate with the Unicorn. The Rosette Nebula, NGC 2237 (Plate 54) in particular, a gigantic 80′ ×60′ wreath-shaped cloud of nebulosity (appropriate for the time of year when this area of the sky becomes prominent), is a legendary object. It resembles the Helix planetary nebula in Aquarius visually, but the Rosette is not a planetary. It's an emission nebula, with the center having been "hollowed out" by the stellar winds of the hot young stars being born in its donut hole.

The Rosette can be difficult to find despite its large size. If you don't have go-to scope, the most expeditious way to locate it is to draw a line from Gamma Orionis, Bellatrix, through Alpha, Betelgeuse, and on for another 8°. This may seem to be a leap into the dark, since you'll see few stars in your target area, but your finder will easily show the open cluster, NGC 2244, the group of brilliant young stars in the center of the Rosette, as a fuzzy patch, so finding will be easier than your star atlas made it look. Not only will the cluster be detectable in your finder, this 24′ across beauty may even be resolved. In the main scope, you'll see a little dipper-like asterism of bright blue–white stars.

The Rosette Nebula itself is often considered a seriously difficult object by visual observers, even those blessed with dark skies. But it's really not, not even in the city, if you remember that it requires two "Fs"—field and filters. Its very large size (it's so big that, in addition to NGC 2237, sections of it also bear the NGC numbers 2238, 2239, and 2246) means long focal length scopes can't help but spread out its light so much that it merges with the sky background. At the proper focal length and magnification it becomes visible in amazingly small instruments. I've even seen the Rosette in my (filtered) 60-mm ETX at 12×.

Of course, going to very low magnifications and wide fields in the city makes the sky background insanely bright. Solution? In this case, an OIII filter. An OIII does an amazing job on this object, and is really required if you are to have a prayer of seeing it in the city. Optimum scope for the Rosette? For me, my 12.5" ƒ/4.8 Dobsonian equipped with a long focal length wide-field 38-mm Plossl and, of course, an OIII. This delivered sufficient light gathering power along with a nice wide-field. Actually, my 8-inch ƒ/5 can do almost as well, delivering enough light, but also able to reach even lower powers with my long-focal-length eyepieces.

What can you expect to see when you find the Rosette? Dim traces of nebulosity scattered across your field, some brighter, some dimmer. The central hole is hard to discern, and, even with the OIII in place, you may find that putting the cluster outside the field of view makes the nebulosity easier to see.

NGC 2264

Talk about "legendary!" The area around the bright variable star S Monocerotis, is that, in spades. Not so much because of the beautiful open cluster, NGC 2264, The Christmas Tree Cluster, but because of the delicate veils of light and dark nebulosity that enwrap this whole area. Foremost among these clouds is the dark Cone Nebula. In the city? Forget it. Even in the country the Cone is an object for very large scopes and talented, experienced observers. But you *will* see the Christmas Tree, a group of bright stars stretching 20′ from S Monocerotis. S also makes the cluster a snap to find. S Monocerotis and its cluster lie 5° north and slightly east of the Rosette. The cluster and S (a variable star that ranges from magnitude 4.2–4.6) are easily apparent in your finder if you just slew north from the Rosette area. Why "Christmas Tree?" That's just what the cluster looks like, the outline of a squat Christmas tree with S Monocerotis forming the "base."

NGC 2261

There *is* another nebula in Monoceros that you can see from the city, one that's considerably less of a challenge than the Rosette. It's small at 2′×1.7′ and bright at magnitude 4 (usually). If not exactly *discovered* by the famous American astronomer, Edwin Hubble, it was at least given considerable study by him and now bears his name. Hubble's Variable Nebula, NGC 2261, is, like M78, a spot of reflection nebulosity— but with a difference. The basic shape of this cloud is triangular or comet shaped, with a star buried in what would be the comet's "head." This star, R Monocerotis, is engaged in blowing off its outer layers, and it is these layers and the interaction between them and the dust and gas already in the area that make the nebula variable. The presence of dark clouds of gas moving from the star and through the clouds, casting obscuring shadows and occasionally blocking some of the star's light cause the nebula to vary fairly dramatically in brightness—sometimes over mere days.

Hubble's Variable Nebula is conveniently located 2° southwest of the Rosette, and is easy to pick up in a medium-power eyepiece if you're careful with your sweeping. A detailed chart of the Rosette area will be a big help. You're looking for a "hazy" star, that, when examined at high power, reveals a trailing, triangular wisp of nebulosity.

Once you've found this nebula, consider making a detailed drawing of it in hopes of detecting changes in it later. Using high magnification and coming back to Hubble frequently over the months it's visible can reveal subtle changes in the nebula to patient observers, even those observing from urban areas—it's that bright. Since this is a reflection nebula, if one composed mainly of gas being thrown off by a sickly star rather than of dust and gas in a star-forming region, light-pollution filters don't seem to help. You'll have to rely on higher magnification and waiting for this object to culminate to provide the best view under city lights.

Tonight's Double: Adhara, Epsilon Canis Majoris

Way down south of Sirius, the brilliant Dog Star in Canis Major, is the southernmost of the two "rear feet" of the big dog, Adhara. This bright magnitude 1.5 beacon is an outstanding double. The white primary and blue secondary are unequal in brightness at magnitudes 1.5 and 7.4, and are somewhat close to each other at 7.5″, so an 8-inch telescope is helpful for an easy split, and a 6-inch scope is often the lower limit if you want to see two completely separate suns under less than perfect seeing conditions. Naturally, the far southern declination of Epsilon, −28° 58′, can cause problems for some Northern Hemisphere observers, but this lovely star should be resolvable from most mid northern latitudes if you wait for culmination.

> It is shiveringly cold and I'm as chilled as it's only possible to be when standing nearly stock-still for hours out in the middle of an observing field (or backyard). The north wind that brought this evening's sparkling clear look at the Milky Way is having its revenge on me. But I'm hooked on the winter Milky Way and the multitudinous clusters and nebulae stretching all the way from Canis Major and Puppis to Orion and Auriga. So it goes, night after night, season after season, year after year. The skies are comfortably unchanging, but they are also forever full of new wonders for me to discover. Even those objects I've seen a hundred times before are still capable of delighting me with their well-remembered but always fresh beauty. On nights like this, I laugh at the city lights as I lift off for the great and wondrous beyond.

Tour 4
Winter's Eskimo

If you were weary of open clusters after finishing the Cassiopeia expedition in the last chapter, I hope you've recovered. In the city, especially in the winter, open star clusters are *the* prime object for the City Lights astronomer. In December the summer stars are gone by the time it's good and dark, and even the constellations of fall are beginning to move to the west by the mid evening. The summer sky was full of globular clusters, and our trek through the autumn heavens also turned up a few. But in deep winter, they are *gone*, with only pale and puny M79 in Lepus present to represent their class. Galaxies? There are a few visible in the winter constellations, but not the endless fuzzballs of autumn and spring.

Forget galaxies and globs for a while. The winter and the Milky Way are a place of open star clusters and nebulae. Revel in them. We can observe these clusters with ease in the worst sky glow. The problem is sifting through long object lists to identify those that are worth a hoot. Too many paltry clumpings barely distinguishable from background star fields can turn off even the most committed galactic cluster fan. Look for rich small-to-medium-sized groups. Learn to decode the Dreyer descriptions, and rely on references like *The Night Sky Observer's Guide* and *Burnham's Celestial Handbook* to help you pick through the winter constellations in search of rewarding destinations.

We'll make stops at some open clusters tonight that can put anything in the summer sky to shame, but, as I've said again and again, the way to deep sky enjoyment is variety and contrast. "Variety" where winter urban observing is concerned leaves you with nebulae in addition to galactic clusters, but most of the season's emission nebulae, once you get beyond the Great Nebula and M78, are frustrating beyond belief for urban observers. Don't spend night after night, as I did as a teenager, searching for the Horsehead Nebula, IC 434, with a 4-inch telescope. Like me, you'll learn a lot about the sky in the Orion area, but you certainly won't see even a trace of old Horsey! What you should do is turn your focus from diffuse emission and reflection nebulae to planetaries. But you will only find one truly outstanding winter planetary (for urban sky watchers) on the Messier list, M76, the Little Dumbbell in Perseus, which we've already visited. What to do? *Get over* the Messier list.

As a young astronomer, I longed for the dark skies of winter, but usually found myself puzzled and feeling a little let down once the first couple of weeks with Orion had passed. What next? The Messier clusters in Auriga, M36, M37, and M38 are delightful, but even they become a little old if they're *all* you look at night after night. What it took for me to break out of the observing doldrums was resolving to sit down with a star atlas and some reference books to find a way to break the Messier habit.

With some study and some exploratory observing, I soon realized that there is a tremendous wealth of winter clusters and planetary nebulae in the NGC catalog, and that many of these are bright and beautiful even in small city scopes. I could go on and on about the forgotten wonders I discovered. "Forgotten" in that many visual observers ignore all but the most well-known examples of these types of objects in the NGC. The whole area from Perseus to Orion holds an abundance of interesting attractions for the traveler willing to stray off the well-beaten path. Tonight, to get you started, we'll visit a series of objects that are so impressive you'll be forever puzzled as to why old Chuck Messier missed them.

NGC 869 and 884

Since Perseus is well positioned for observing at an early hour at this time of the year, we'll begin within its borders with the famous Double Cluster, NGC 869 and 884. Though not in the Messier, this object is, admittedly, not among the "forgotten" clusters of the NGC. All reasonably experienced deep sky observers know about the Double Cluster, and it is not to be passed by on any winters evening.

A large part of enjoying the Double Cluster lies in knowing *how* to observe it. As the name indicates, this is two star clusters, open star clusters, NGC 869 and 884, magnitudes 4.3 and 4.4, respectively, the centers of which are separated by only 30'. Each is beautiful on its own, but put both in the same eyepiece field, and you have one of the most amazing objects in the sky. For full enjoyment, you'll need to use an instrument that will accommodate both and leave a little space around them. This can be a problem—your average *f*/10 SCT won't give you the at least 2° expanse you need to take in all of the scattered cluster members. However, you do it—focal reducer, long focal length eyepiece, short tube refractor, or binoculars—I urge you to view this pair the way they were meant to be seen, in one field, like the sprawling lights of two huge adjoining cities set on a dark plain. For me they looked amazingly like Dallas and Fort Worth, Texas as seen from a red-eye flight.

Finding the Double Cluster is easy enough. One way to locate the pair is to put Magnitude 3.76 Eta Persei in the finder crosshairs and scan 4° northwest in the direction of magnitude 2.68 Delta Cassiopeiae. An elongated haze patch will jump right out of your finder at you when the Double Cluster comes into view. When the clusters are nearing culmination, you may even be able to catch them with the naked eye; certainly they are easily visible as a double misty patch without optical aid from medium-dark suburban sites. When you have the two in the field of the main scope, spend some time absorbing their majesty. Then start comparing them and picking out details.

I found my best views came with my 8-inch *f*/5 Chinese reflector and a 38-mm Plossl eyepiece with a 60° apparent field. This provided a 75' wide expanse of sky, and while not *quite* enough to take in every straggling star, it framed the pair nicely. Going lower in power would've tended to amplify the sky glow beyond what can be tolerated. A more expensive wide-field eyepiece would, of course, go even further toward hitting the magic 2° field size perfect for the Double Cluster. I could have also gone to a smaller, shorter focal length refractor, but the 8-inch *f*/5 seemed made for this object, easily taking in the rich centers of both clusters and providing plenty of light gathering power. In the 38-mm eyepiece, the westernmost cluster, NGC 869 was considerably richer, more condensed and generally flashier than its companion, but both were outright spectacles:

> The Double Cluster fits easily into the field of my inexpensive 38-mm medium wide-field 2-inch format Plossl eyepiece (26×) in my 8-inch f/5 Synta Newtonian. The eyepiece field is somewhat gray at this low power, but both clusters are very lovely and extremely

rich. Going to higher power with another imported eyepiece, a 26-mm wide-field design, improves the view somewhat, even if some of the cluster members are placed outside the field, since the background sky is considerably darker at higher power. I quickly went back to the 38-mm eyepiece, however, as this object just has to have wide field to be really enjoyed. The grayness of the background is not that noticeable due to the brilliance of the cluster stars. They almost seem to be bright enough to make me lose my dark adaptation!

In the deep sky game, you get used to attractive illusions. *That* planetary nebula is not *really* a member of *that* star cluster, and *that* cluster is not *really* involved in *that* nebula. They are merely along the same line of sight, put together in the eyepiece by our unique perspective. In this case, though, the two clusters of the Double Cluster are actually almost as intertwined as they appear to be in our telescopes. They are fairly close in space, only a few hundred light years apart. Both are roughly 7,000 light years distant from Earth, and are part of the great Perseus OB-1 association of young stars.

NGC 1528

Everybody's heard of the Double Cluster, but who knows NGC 1528? It's a shame that reciting its catalog number will result in blank stares from most amateur astronomers. It's a remarkably beautiful open star cluster. I didn't have any difficulty at all in locating this object, either, since it lies only 1° 37′ east of the prominent star Lambda Persei in northeastern Perseus. NGC 1528 forms a near-right triangle with Lambda and nearby Mu. Lambda and Mu are not terrifically bright (both are about magnitude 4.5), but they may even be naked eye visible when Perseus is high in the sky on dark winter nights. Using these two stars as your guideposts, sweep carefully, and you will soon have this cluster centered. NGC 1528 has a combined magnitude of 6.4 and is 24′ across, so a lower power is demanded both when searching for and when observing this object.

I think you'll be as surprised as I was at how nice this cluster turns out to be. Since the NGC galaxies are *usually* much dimmer than the Messier galaxies, as a new observer I was prepared for the NGC open clusters to also be inferior to their Messier counterparts. But that is not always the case, and certainly not this time. This object seems superior to such famous Messier clusters as M29 and M39 in Cygnus. Owing to its medium size, NGC 1528 will fit nicely in the field of a medium-power eyepiece, making a very pretty picture. While it contains about 165 stars photographically, if your conditions are similar to mine, you'll probably be able to make out, at most, about 20–25 suns. But this is still a remarkable object. An extract from my observer's log for NGC 1528 reads:

> Quite a little gem in the 8-inch f/5 Newtonian. Fairly rich, and dominated by a 'U' shaped asterism. Around 30 cluster members are seen along with quite a bit of haze suggesting more unresolved stars. Best seen in the 22-mm Panoptic, though it would probably do well with higher magnification. I'm very impressed by this cluster.

NGC 1528 is located 2,600 light years from Earth, and in the Trumpler classification scheme for open clusters, it is a "detached, medium-rich cluster with a medium range of star brightness."

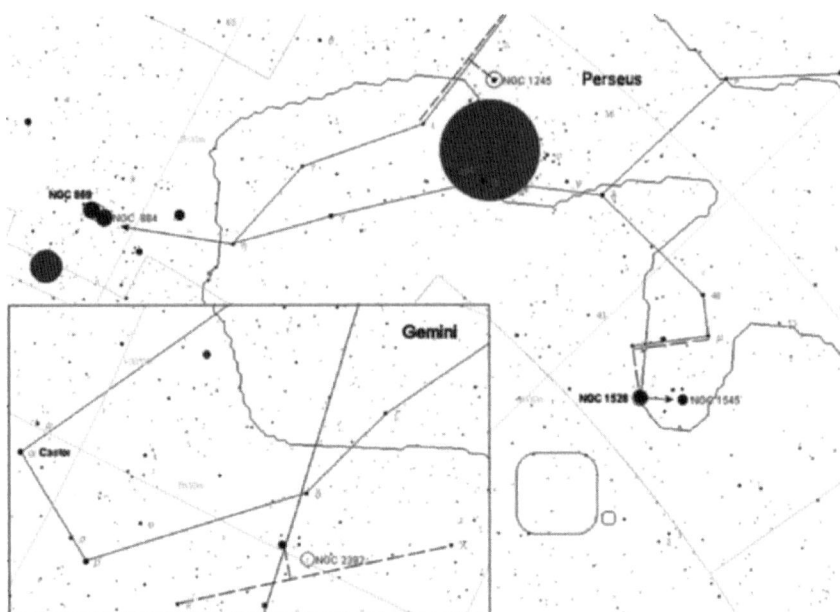

Figure 9.8. Gemini and Perseus.

The Eskimo Nebula, NGC 2392

Much as I love the Double Cluster and enjoy nice surprises like NGC 1528, there's no doubt that the evening's final *feature* stop is the "best" of the night—in my opinion, anyway. Let's turn to the celestial twins, Gemini. NGC 2392, a real deep sky wonder if ever there was one, is another amazing showpiece that Messier missed. This is the famous Eskimo Nebula.

NGC 2392 is even easier to find than NGC 1528. It lies almost exactly at the center of a line drawn between Lambda and Kappa Geminorum, and about 2° east of Delta. Since this a small object (40" in diameter), you may want to use a higher power than usual when searching for it. Work slowly, keeping the chart in Figure 9.8 oriented to match the view in your finder. Its magnitude 10.0 disk is obviously nonstellar even at low magnifications.

Once you've located the Eskimo, the first thing you'll notice is its surprisingly obvious blue–gray color. You'll probably also be amazed at just how bright it seems. This object will definitely take high magnification, so use all the power your telescope and sky will permit. The 10th magnitude central star of NGC 2392 is readily visible, and in a small-to-medium-sized telescope the nebulosity winks in and out as you avert your vision, just like Cygnus' Blinking Planetary. Although the nebulous disk surrounding the central star is bright, there were only the very *faintest* hints of detail in a 4" scope. I didn't really see *much* more with the 12.5-inch scope in the

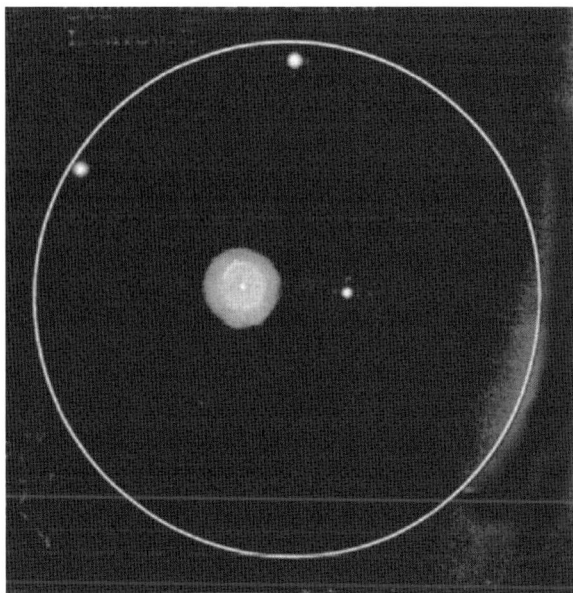

Figure 9.9. The Eskimo shows hints of detail in a 12-inch scope.

city, just enough faint indications of internal structure to tantalize me; just enough to record in the drawing in Figure 9.5. Why is this planetary called the "Eskimo"? Look at a long-exposure photograph of it and you'll see that the combination of an inner, mottled, ring of nebulosity with a fuzzy outer ring looks amazingly like a human face in a hooded parka (see Plate 55). This detail doesn't come easy for visual observers, but some of these features are visible in large amateur scopes under dark skies using very high magnifications.

> I have a very nice view of the Eskimo in the 12.5-inch Dobsonian. In moments of good seeing, detail seems to flash into view at about 300×. But it's gone before I can decide exactly what I've seen. I can definitely make out that there is an inner and an outer ring of nebulosity. This is one of those times when drawing is the best way of "seeing" an object. When I spot a detail, I immediately record it. It vanishes, I wait for the next flash of detail, record that, and so on until I have a nice "time exposure" sketch.

In the drawing, in Figure 9.9, it's easy to see that, despite all my work, I couldn't coax a lot out of the Eskimo. Can I do better? Maybe. Steadier skies than are usually present in winter are often a huge help with small deep sky objects (DSOs) at high magnification. I'll come back to this planetary nebula with sketchbook and determination on a "planetary seeing" night when I'd normally be after Jupiter or Saturn.

NGC 2392 is *thought* to be *about* 3,000 light years away. Given its rate of expansion, this star-corpse is *believed* to be one of the youngest of the planetary nebulae, with a tentative age of less than 2,000 years. The Eskimo was first observed by William Herschel in 1787.

More? Thanks to my usually balmy wintertime temperatures, I can often go all night, cruising up and down the Milky Way's downtown strip. If you want more, look to these standouts.

NGC 1245

This is one of Perseus' better DSOs, and that is obviously saying a lot. A 10' in diameter open cluster, it is fairly rich, showing off at least 15–20 suns at medium power in the 8-inch *f*/5. At a magnitude of 8.4, it's dimmer than Perseus' other standouts, so catch it when it's near culmination, and use medium magnifications to pull out as many dim members as you can. As noted earlier, the Perseus area is so littered with clusters and asterisms that *look* like clusters that it can be difficult to find exactly the group you're looking for. For once, though, I didn't have to slew around wondering where my cluster was. I positioned the scope halfway along a line drawn between Iota and Kappa Persei, and then moved the scope just under a degree—51'—to the East. I couldn't pick up NGC 1245 in the finder, but when I moved to the main scope, it was awaiting me in a medium-low power eyepiece, showing a distinctive "W" shaped asterism formed by its brightest stars.

NGC 1545

Another interesting open cluster, this is a scattered but interesting group 18' diameter, glowing at magnitude 4.6 at the feet of Perseus. Like NGC 1245, it is easy to locate, since it forms an equilateral triangle with easily seen Lambda and 48 Persei. Another pointer to the group is a pair of magnitude 5 stars 2° East of Lambda. The cluster is 26', one medium-power field, further to the east. While not as eye-catching as some of the area's agglomerations, this is a good "meat and potatoes" object. Three bright stars form a small triangle at the group's heart, and are surrounded by perhaps 20–25 obvious cluster members.

Tonight's Double Star: Castor, Alpha Geminorum

The most novice of sky watchers is familiar with Castor, Pollux's twin in the classical constellation, Gemini. But fewer amateurs than you'd think are aware that the bright Alpha star is a good, if close, double. Yes, Castor is actually two stars, a magnitude 1.9 primary and a magnitude 2.2 companion. The two are at a frighteningly close separation of 2.2", but their similar magnitudes make them easy in small scopes. I've routinely split this pair of pure white stars with my 80-mm *f*/5 refractor at high magnification under good conditions. If you find it difficult to split Castor, make sure you're not really looking at Pollux. Castor is the *westernmost* of the two bright stars. If you know this area of the sky, a good way to remember which star is which is, "Castor is close to Capella, Pollux is in proximity to Procyon."

When you finish packing-up your beloved telescope and head-in from a late winter night's cold, stop for a moment and ponder what we've seen tonight. These open clusters and the planetary nebula represent the Alpha and Omega in the lives of the stars—birth and death. Our clusters are maternity wards full of young stars. The nebula is an aged sun on its deathbed and has thrown off its outer layers and begun a long, long decline. Even the stars do not forever endure. But for ephemeral creatures like us they remain as unchanged and beautifully new as on the night we took that first wondering look up.

Tour 5
One for the Road

As winter grows old, a constellation begins to dominate the heavens even more strongly than spectacular Orion. Obviously, the main reason for Canis Major's prominence is magnitude −1.5 Sirius, the brightest star in Earth's sky. The basic stick-figure shape of the Big Dog, an inverted "Y," is easily identifiable, but unmemorable. It's the blazingly bright blue–white sapphire of Sirius that makes up for the lack-luster shape of its home constellation. Sirius is so bright that when it's near the horizon, flashing every color of the rainbow due to the thick atmosphere there, it often results in frantic UFO reports from the naïve and gullible.

You'd think a mid-sized constellation—it covers 380 square degrees of sky—lying near the winter Milky Way would be loaded with deep sky wonders. Sadly this is not the case if your definition of "wonder" is "a bright Messier object." There's only one Messier within Canis Major's border, the large but outstanding open cluster M41. There are a couple of other galactic clusters, M46 and M47, which we've already visited, in nearby Puppis, but not much else of the showpiece class is to be found in the vicinity of the Big Dog. A little digging in the star atlas did turn up some additional objects of interesting types in the area—including a two diffuse nebulae and a galaxy for us to visit—but these are subtle objects, so be prepared to do some real work.

M41

M41 seems more than deserving of its inclusion in the Messier catalog. It's bright at magnitude 4.6 and very easy to find in the relatively barren star fields near the western end of Canis Major. Unfortunately, it is large, 38′ across its major axis, and a little thin star-wise when compared to the glorious clusters in Auriga, for example. It *is* magnificent in binoculars, with my inexpensive Chinese 15×70s providing an unforgettable view. In these glasses, it's easily resolved into myriad stars, and appears as a compact and elongated object reminiscent of a globular cluster in a telescope. It looks so fantastic in binoculars or my Short Tube 80 refractor at low power that I'm always disappointed when switching to the main scope, and seeing this cluster revealed as a nice but rather pedestrian assemblage. Bottom line? For best effect, keep the magnification down and the field wide.

It can be a little difficult to find the right magnification and the right field diameter to show this M object to best advantage. But finding this object couldn't be easier. The chart in Figure 9.10 shows it sitting pretty 4° almost due south of Sirius, forming a right angle with Sirius and Beta Canis Majoris, magnitude 1.98 Murzim.

In the eyepiece of the 8-inch *f*/5 Newtonian telescope, M41 is revealed as a medium rich and somewhat shapeless group. In the city, expect to see about 30–40 stars with ease (under dark skies even a 3-inch scope will reveal 50 or more stars here). This elongated, flattened pattern of bright stars is supplemented and enhanced by scads of dimmer suns spangled across the field in random fashion that begin to show themselves

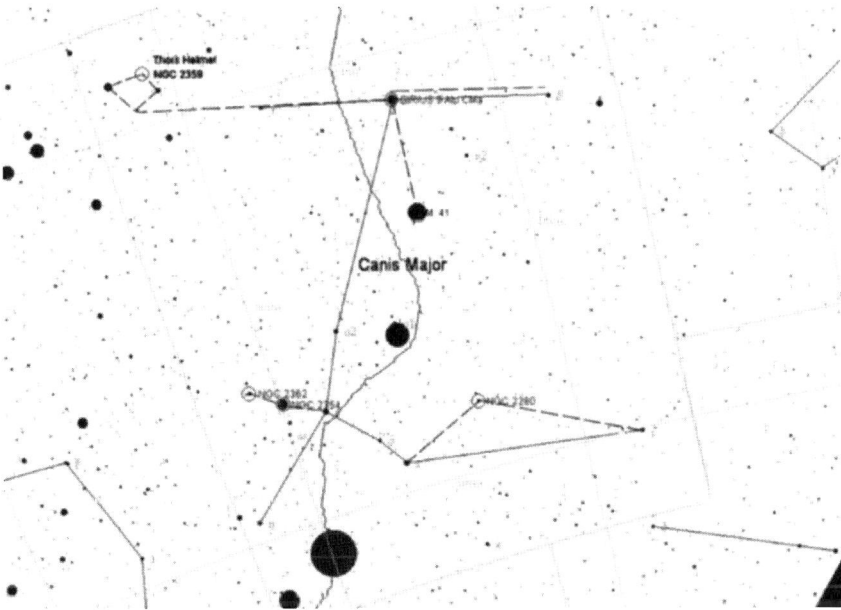

Figure 9.10. Canis Major chart.

with increased magnification. Upping the power does cause the cluster to lose impact by eliminating the dark sky around it that frames it at low power. But increasing the magnification reveals many stars that were invisible at lower power. I always try to examine open clusters at a variety of magnifications, even if the smaller fields of short focal length oculars "destroy" the beauty of the larger groups. You'll see several bright stars involved in M41, with 12 Canis Majoris, which lies just 20′ to the southeast of the cluster's heart, being most impressive at magnitude 6.0.

Yes, I was happiest with M41 in large-aperture binoculars or a wide field refractor, but I did enjoy viewing it in the 8-inch *f*/5:

> This outstanding Messier cluster shows off at least 25 members in the 8-inch with a 35-mm Panoptic at 28×. The Short Tube 80-mm refractor shows off almost as many of stars, but provides a much more impressive view at 20×. The most distinguishing characteristic of the cluster is a shockingly orange–red magnitude 9 star near the center of the group. This is made even more impressive by its contrast with its sister cluster stars, most of which are blue–white in color. A longer look at the cluster reveals numerous oddly curving chains of stars and hints of a background haze of unresolved suns.

Under dark skies, M41 can be a naked eye object, appearing to be almost the size of the Full Moon, at least if you live far south enough in latitude to get it out of the trashy air at the horizon. M41 was first catalogued in the 17th century, well before Messier laid eyes on it, but this cluster may have been seen by the ancient Greeks long before that, and, if so, was one of the few DSOs known to them (M44, the Beehive, being another). Current thinking places M41 at 2,300 light years, making it approximately 25 light years in size.

Thor's Helmet, NGC 2359

With NGC 2359, we go from easy and obvious and pedestrian to hard, challenging, and interesting. Canis Major, trotting along beside his master, Orion, and splashing his paws in the edge of the Milky Way's stream, is not as rich in emission nebulae as you'd expect; certainly he's not as blessed with these objects as the Hunter. There is at least one nice example of this class of DSO in Canis Major, NGC 2359, The Thor's Helmet Nebula (also occasionally referred to as the "Florida" Nebula).

Thor is an unusual and attractive object that definitely deserves a look-see if you're up to his challenge. Be prepared to throw as much aperture as you can muster at it. In my skies it took the Nexstar C11 to pull it out. In part, this is due to the southerly declination of this object, −13° 14′, but also because of its intrinsic dimness. NGC 2359 is 10′ in extent, fairly large, and its surface brightness is resultantly low. On typically hazy (for my location) late winter nights, it definitely required a UHC filter for detection. On an outstanding evening, this object *should* be visible in an 8-inch scope—I've thought I've seen it, barely, in my UHC filtered 8-inch *f*/5 on fair nights—but in the city it is most often a nebula for 10-inch telescopes and above. In addition to aperture, an OIII or UHC filter is mandatory.

While not overly difficult to locate, there aren't any bright nearby stars to direct you to Thor, so search carefully using a low-power, wide-field ocular with a UHC or OIII filter in place. Start at Sirius, and draw a line through Gamma Canis Majoris, a distance of 4° 35′. Extend this line an equal distance out into the empty space and stop. The nebula lies approximately 45′ northwest of your stopping place.

I knew that if I used my computerized, go-to-equipped Nexstar 11, finding Thor's Helmet obviously wouldn't be a problem, but I was skeptical as to whether I'd see anything once the telescope landed me on the correct field. Even under dark skies, this is not always an easy nebula in a 10–12-inch SCT. I was very pleased, then, to be able to see Thor's Helmet with direct vision without even trying very hard on an average evening:

> Thor's Helmet is surprisingly visible under relatively poor, hazy conditions. Despite the light pollution, it was easy in a 35-mm Panoptic at 80× when I used a UHC filter in conjunction with the eyepiece. Without the filter, the nebula is seemingly invisible—I didn't think I could detect it, anyway. It does appear that I'm seeing almost the full 10′ swath of nebulosity, but the "helmet wings" seen under dark skies are not visible, or at least only barely. Most of the time, this is just a dim, elongated patch slightly brighter than the sky background, but I occasionally convince myself that I can detect one curving, elongated swath of nebulosity after long observation with a dark hood over my head.

The Thor's Helmet Nebula (Plate 56) is a type of nebula completely different from those we've visited most of in this series of tours. It's not an emission nebula or a reflection nebula, nor is it the remains of a supernova or a Solar mass star. Thor's Helmet is a type of nebula associated with Wolf-Rayet stars. Wolf-Rayet stars are very hot (up to 50,000 kelvin) and very large (up to 20 Solar masses) stars that are well down the evolutionary path to supernovae-hood and are expelling their outer layers. The nebula is the remains of the stellar atmosphere blown off by a nondescript-looking "central star." At some point in the distant future, Thor's associated star is likely to burst into supernova glory in the star fields of Canis Major. Another example of this phenomenon is the (in)famous Bubble Nebula in Cassiopeia, which is far more difficult to detect in city or country than Thor's magnificent Helmet.

Tonight's Double Star: Sirius, Alpha Canis Majoris

We can hardly leave Canis Major without taking a look at his stellar attraction, the brightest star of them all. Normally, looking at a bright star in a telescope is not a very interesting experience. In a scope, a brilliant star is just that, a bright point of light, maybe dancing around as seeing changes. Children do seem fascinated by the appearance of Sirius or any bright star in the eyepiece, but there's not much here for amateur astronomers. Not until recently. As of this writing, 2005, the Pup has come into view.

Sirius is possessed of a companion star, Sirius B, the "Pup," that's fairly easily visible at certain times. On average, the Pup lies at a distance from its parent star similar to the planet Uranus' distance from our Sun. The orbit of Sirius is quite a bit more eccentric—elliptical—than that of Uranus, however. From our perspective the Pup can be up to 11″ from Sirius, but this can dwindle down to 3″ at the other "side" of the Pup's orbit. Certainly, 3″ is not terribly tight as close double stars go, so the Pup should be easy to resolve, right? No. Not hardly. The problem is the huge difference in magnitudes. The Pup is at magnitude 8.5, while Sirius is blazing away at magnitude *minus* 1.5.

This immense difference in brightness makes the companion star hard to see, even at maximum separation (the Pup will reach its greatest distance from Sirius in 2005 and will begin to slowly move closer thereafter). At or close to the Pup's greatest "elongation," the careful observer should be able to resolve it with a 10-inch scope at high power—300× and above. Sirius B can probably be detected in considerably smaller telescopes, but I'm finding that the extra aperture makes detection much more certain. With the C11 at 350×, the Pup is unmistakable as a little spark of light barely separate from gaudy Sirius' "rays." To be positive I wasn't fooling myself, I printed a finder chart for the Pup oriented to match the view produced by my scope (*Skytools 2* allowed me to produce a chart for mid-2005, Figure 9.11, in just a couple of minutes). Frankly, if you're not a double star fan, the Pup isn't overly impressive. Not unless you know the fascinating and slightly silly story surrounding it.

The discovery of Sirius B is straightforward. Famous American telescope maker Alvan Clark spotted it as he was testing a brand new telescope, an 18.5-inch refractor he was installing at Dearborn Observatory in Michigan in the U.S. on January 31, 1862. That it took until the mid 19th century to discover the Pup is a little surprising, but not astonishing. While large scopes had been in use since the time of William Herschel in the 18th century, the *quality* of telescope optics didn't begin to catch up until the 19th century, when quality was dramatically exemplified by Clark's exquisite refractor objectives.

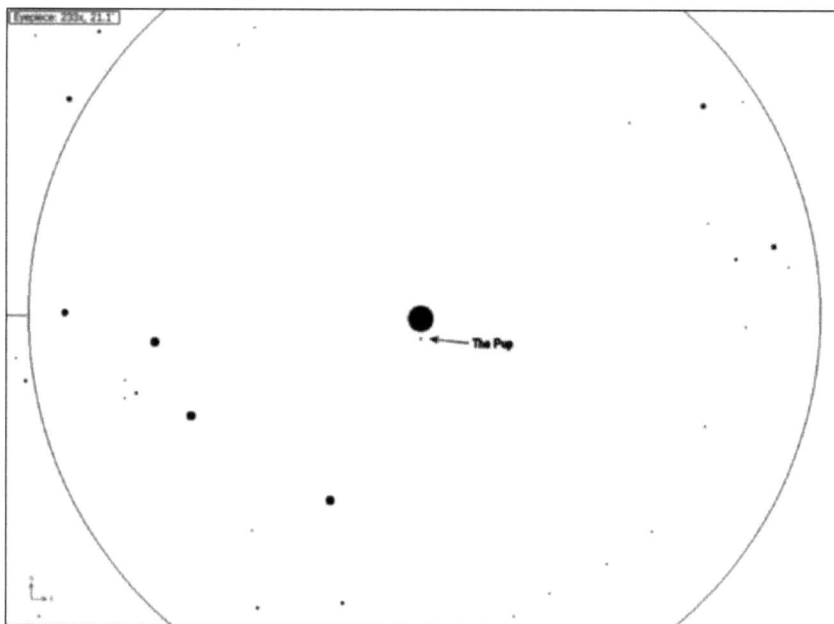

Figure 9.11. *Skytools 2* finder chart for Sirius and his Pup.

The interesting part of the story didn't begin until nearly a century later, in the 1940s. At that time, two French anthropologists were studying a West African tribe, the Dogon, which supposedly had little contact with westerners. These people, who lived in what is now Mali, fascinated the scientists with folklore stories that seemed to indicate that they had long been aware of the existence of a dim companion to Sirius. Not only that. They also claimed to know that Sirius B is a very dense white, dwarf star, something that was fresh knowledge to astronomers of the late 1940s. *And* the Dogon appeared to know that the Pup orbited Sirius over a period of 50 years. How did they know these things? How could they *possibly* know? The Dogon claimed to have been visited by extraterrestrial voyagers some 10,000 years previously, who gifted them with their knowledge of the Pup.

What a story. It's easy to see why the tale of the Dogon and the Pup has become beloved of UFO buffs and *X Files* fans. But what's the ground truth? How to explain the tribe's apparent knowledge of advanced astronomical data? The first theory to be offered was bandied about by people in the anthropology community who didn't have much knowledge of observational astronomy. Maybe, they theorized, the dark skies of West Africa enabled the Dogon to see Sirius B without optical aid. This "explanation" seems laughable. Even under the darkest skies it's unlikely that *any* human will see down to magnitude 8.5, and even if someone could, the close proximity of Sirius would make detection totally impossible, even at greatest elongation of B from the main star.

Another possibility is *cultural contamination*. During the 1920s and 1930s there was a great deal of travel to West Africa by westerners of all types, not just scientists, at a time when there was much speculation in the European press about the nature of Sirius B as a dwarf star. It's not clear that the anthropologists were really the first Europeans to have encountered the Dogon. It's entirely possible that other western visitors sitting outside with their hosts at night might have related the news of Sirius' strange nature to impress the tribesmen. This cultural contamination mixed with preexisting folklore beliefs of the Dogon seems the best explanation for the mystery. The Dog Star/Dogon tale *does* make a good story though, and is not completely and convincingly explained to my satisfaction yet.

Sirius, Alpha Canis Majoris, is a blue–white A1 spectral class star located 8.6 light years from Earth. It has a mass about 1.5 times that of the Sun, but its brilliance is due to its relative closeness, as it is not a giant star. It is famous in history due to its association with the flooding of the Nile (in ancient Egypt, Sirius rising with the Sun meant it was time for the Nile to leave its banks), important to early Egyptian agriculture. Its moniker, "Dog Star," also comes from the Egyptians and their association of it with their dog-headed god, Osirus.

Elsewhere in the Kennel

There's no question that Canis Major is the land of open clusters. A look at a detailed star chart will reveal hordes of them. Most of these NGC objects are fairly lackluster, but there are a couple of nice exceptions.

NGC 2354

A magnitude 8.9, 20′ diameter open cluster located 1° 29′ northeast of Delta Canis Majoris. NGC 2354 is not overly rich, but its interesting shape makes it worth a look. The center of the cluster is marked by a relatively bright magnitude 9 star that is surrounded by an almost complete ring of medium-bright cluster stars. This strange asterism almost looks artificial.

NGC 2362

Slewing the telescope 1° 14′ farther to the northeast from NGC 2354 brings us to NGC 2362. Finding this one is as easy as falling off a log, since it's involved with bright magnitude 4.39 Tau Canis Majoris. There are as many as 60 stars visible here, though I see less than half that many from my light-polluted stomping grounds. Roughly triangular in shape, most of the cluster is composed of dim stars. This contrast between Tau and the rest of the group can produce an interesting effect. Tau is known as the "Jumping Spider Star" because a tap on the scope's tube will cause the star to jiggle in a pattern seemingly *different* from the rest of the cluster stars. This strange optical illusion is due, no doubt, to the large brightness difference between Tau and its companions.

NGC 2280

There are a few other types of objects in addition star clusters to be seen in Big Dog's den, Thor's Helmet being one example. If you comb the star fields you can even turn up a number of galaxies. Unfortunately, most of them are *insanely* dim for the City Lights astronomer. You do have a shot at NGC 2280 (Plate 57). At magnitude 11 and $7.0' \times 3.0'$, it is often doable on very good city nights by medium-to-large aperture scopes. Wait for its field to get as high in the sky as it will, and examine the area carefully at medium-high magnification. Don't expect much more than a faint, perhaps vaguely elongated (northwest–southeast) smudge, however. The spot where this shy galaxy lurks is easy to get to, since the bright foot of the dog, Epsilon, and magnitude 3.02 Zeta Canis Majoris (see Figure 9.6) form a flat triangle with the galaxy.

> Sirius, light of the South! His brilliant blue beacon seems to overwhelm the dull glow of sodium arc streetlights. Whether hunting the elusive Pup or searching for the sometimes maddeningly difficult Thor's Helmet Nebula, I've again forgotten I'm observing under "impossible" city skies, and have just had a wonderful time out under my stars.

Canis Major marks the end of our starry road. I hope you've enjoyed this series of deep sky hikes and will go on to plan and take many more on your own. I'm always interested in the urban adventures of my brother and sister amateur astronomers, and would enjoy hearing from you on the "Rod's City Lights" computer mailing list (see Appendix 1 for the URL).

Happy deep sky hunting from city lights!

Internet Resources

Web Sites: Reviews and Resources

- *Adventures in Deep Space* (http://www.angelfire.com/id/jsredshift/). The focus of this site is "challenging observing projects" aimed at dark-site observers, but this excellent collection of pages contains much information on observing the deep sky of interest to City Lights astronomers.
- *The Astronomical League* (http://www.astroleague.org/). The prime attraction is the Observing Clubs section, and especially the Urban Club.
- *Sky Charts* (http://www.stargazing.net/astropc/). The home page for *Cartes du Ciel*, the excellent freeware planetarium program. The charts in this book, are based on the output of this wonderful planetarium program.
- *The NASA Extragalactic Database* (http://nedwww.ipac.caltech.edu/). Aimed largely at professional astronomers, this is also an incredible tool for the amateur, with detailed information on thousands and thousands of galaxies.
- *SEDS Messier Database* (http://www.seds.org/messier/). This online illustrated Messier catalog is one of my most frequently used web sites. In addition to in-depth info on each object, it includes links to other resources on the Messier objects.
- *SEDS Interactive NGC Catalog Online* (http://www.seds.org/~spider/ngc/ngc.htm). SEDS also has the NGC catalog available in a format similar to its online Messier catalog. Not quite as well done or information-rich as the Messier database, this is still a handy and quick way to get the vital statistics of NGC objects.
- *Sketching and Observing the Deep Sky* (http://www.skyrover.net/ds/). Thousands of excellent drawings of deep sky objects (DSOs). I often find drawings of DSOs more indicative of what I'll see through the eyepiece than the best long-exposure images.

The Urban Astronomer's Guide

- *Skyhound* (http://skyhound.com/sh/skyhound.html). This site is produced by Greg Crinklaw, developer of the *Skytools 2* program, and contains information on objects appropriate for observing for the current month. It also contains features and information on deep sky observing of interest to all observers.

Mailing Lists

- *Rod's City Lights* (*Yahoo Group*) (http://groups.yahoo.com/group/rodscitylights/). The author's Yahoo mailing list for discussion of this book and city-based observing in general.
- *Urban Astronomers* (*Yahoo Group*) (http://groups.yahoo.com/group/urban_astronomers). This online forum is not very active, but when there's message traffic it shows great potential.

Digitized Sky Survey Images

The DSS images used throughout this book are here thanks to the kind permission of Palomar Observatory and the Digitized Sky Survey created by the Space Telescope Science Institute (STScI), operated by AURA, Inc. for NASA:

> The Digitized Sky Surveys were produced at the Space Telescope Science Institute under U.S. Government grant NAG W-2166. The images of these surveys are based on photographic data obtained using the Oschin Schmidt Telescope on Palomar Mountain and the UK Schmidt Telescope. The plates were processed into the present compressed digital form with the permission of these institutions.
>
> The National Geographic Society—Palomar Observatory Sky Atlas (POSS-I) was made by the California Institute of Technology with grants from the National Geographic Society.
>
> The Oschin Schmidt Telescope is operated by the California Institute of Technology and Palomar Observatory.

APPENDIX 2

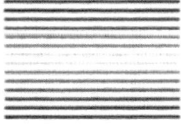

Finding Directions in the Sky

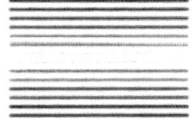

Degrees, Minutes, and Seconds

From our perspective, the sky appears to be a globe just like the globe of the Earth, and can, like the Earth, be divided into 360° of circumference. We define distance in the sky using this system. Object "X" is said, for example, to be 10° from object "Y." How far is a degree in the sky? Luckily, there's a simple measurement system that's been used for ages and which is surprisingly accurate. 1° is the width of your index finger held at arm's length. 10° is the width of your closed fist, again held at arm's length. 20° is the distance from the tip of your thumb to the tip of your little finger on your spread hand. A degree is further divided into 60 arc minutes ('), and a minute is split into 60 arc seconds ("). The Moon is about 30' (minutes) in diameter.

To find objects using the directions given in the tours in this book, it is vital you know the diameter in degrees and/or minutes of your eyepiece and finder fields. Determine the amount of true field your eyepieces show with a particular telescope by dividing an eyepiece's *apparent* field (given by the manufacturer in his specifications for the eyepiece) by the magnification it yields in your instrument. A 25-mm eyepiece with a 50° apparent field that's delivering a magnification of 80× yields a true field of 37.2 min, a little over half a degree:

$$50/80 = 0.62 \text{ degrees}$$
$$60 * 0.62 = 37.2 \text{ minutes}$$

The easiest way to find the size of your finder's true field is by placing a bright star on one edge of its field. Find another prominent star directly across from this one in the field. Referencing a detailed star chart, measure the distance between the two in degrees and minutes (use the declination scale). My telescope's finder just barely includes both Castor and Pollux in one field. Looking at a chart, I can see that makes my finder's field about 4° 30' in diameter.

Right Ascension (RA) and Declination (Dec)

Right ascension (RA) and declination (Dec) are the longitude and latitude of the celestial sphere. Just as it's more convenient to say a city lies at 30° north and 88° west instead of saying that it is 10° south of New York City, it's convenient to be able to describe a location in the sky in exact terms. RA and Dec seem difficult to understand at first, but they are really easy concepts; if you understand Earthly latitude and longitude, you already understand RA and Dec. Declination is measured from the celestial equator, which is at 0° declination, to the poles which are at 90°. Degrees north of the celestial equator is expressed as a positive (+) value (if there's no sign given, the Dec is assumed to be north), while degrees south of this imaginary line is expressed as a negative (−) declination. The north celestial pole is at a declination of +90°; the south celestial pole is −90° declination.

On Earth, Longitude is measured from Greenwich, England. In the sky, some arbitrary point had to be chosen for celestial longitude to "begin." That place in the sky is *The First Point of Aries*, the Vernal Equinox, the spot where the Sun crosses the celestial equator on its way back North. RA increases to the east and, unlike Earthly longitude, is expressed in *hours, minutes, and seconds* rather than degrees, minutes, and seconds. This makes sense because the celestial sphere is "turning" once every 24 h (approximately). Point your scope at a spot in the sky, turn off the drive, come back in an hour, and the scope will now be focused on a spot *one hour of RA east* of its original point. The celestial globe has turned one hour of RA westward. One hour of RA is equivalent to 15° in the sky—the celestial globe is rotating at a rate of 15° per hour.

Directions in the Sky

Often, when novice astronomers are trying to describe an object's position in the sky they will say something like, "It's just above and to the right of that big star." We all occasionally use "up," "down," "right," and "left" to describe sky positions, but this is an awkward and ambiguous way to describe locations. Much better is to use compass directions. If I say a galaxy is 5° east of Spica, you have a pretty good idea of exactly where it is in the sky, no matter what Spica's current position. To get a grip on compass directions in a telescope eyepiece, forget about terrestrial map directions and think only in terms of the sky. North is in the direction of Polaris. South is opposite. East is always at a right angle to north. West is the opposite direction.

To locate objects, you need to know what the compass directions are in your finder and in your main scope. In either telescope or finder field of view, no matter how the view is inverted or reversed, east is the direction where stars *enter* the field with the telescope's clock drive turned off (or when you move your scope toward the eastern horizon). West is where they *leave* the field with the drive off. Move the telescope west, and the stars will *enter* on the west side of the field and *leave* on the east. Slew your scope toward Polaris, and stars will *enter* the field from the north. South is the direction where stars *leave* the field when you move north. Nudge the telescope to the south, away from Polaris, and objects will *enter* on the southern side of the field and *leave* on the north. *Practice determining compass directions in the eyepiece field until you are as comfortable as possible with this concept.*

Object Classifications and Descriptions

Globular Clusters

Globular clusters, ancient balls of stars orbiting the Milky Way's center, are sorted according to the classification scheme developed by Harlow Shapley and Helen Sawyer Hogg. In this system, they are rated on a 12-step scale from I (very concentrated) to XII (not concentrated). It's common to see Shapley–Sawyer classes written as either Roman numerals or Arabic numbers.

Open Clusters

The Trumpler system for classifying open (galactic) star clusters, groups of newborn stars that reside in the Milky Way's spiral arms, was developed by R.J. Trumpler in the 1930s, and groups open clusters according to following three characteristics: concentration, range in brightness, and richness.

- *Concentration:*
    - I. Detached, strong concentration toward center;
    - II. Detached, weak concentration toward center;
    - III. Detached, no concentration toward center;
    - IV. Not detached from surrounding star field
- *Range in Brightness:*
    1. Small range;
    2. Moderate range;
    3. Large range.

- *Richness:*
    - p. Poor (<50 stars);
    - m. Moderately rich (>50 stars <100 stars);
    - r. Rich (>100 stars).

An open cluster labeled as a "II, 1, p," for example, would be a "detached cluster with weak concentration with a small range in star brightness, which is poor in richness, containing 50 stars or less."

Galaxies

The famous Hubble Type galaxy classification system arranges galaxies, "island universes," according to basic appearance, characteristics of their arms and nuclei, and other features such as flatness in the case of ellipticals.

Elliptical Galaxies (E), E0–E7

E0 is a spherical galaxy, E7 is a highly flattened, almost saucer-shaped elliptical.

- *Other Elliptical Specifiers:*
    - d: Dwarf;
    - c: Supergiant;
    - D: possesses a diffuse halo.

S0 Galaxies are transitional types between spirals and ellipticals. They share some characteristics of both types, but these disk-shaped galaxies are not necessarily evolved from one type or evolving into another.

Spiral Galaxies (S)

Sa: tightly wound arms;
Sb: moderately wound arms;
Sc: loosely wound arms.

Barred Spiral Galaxies (SB). These are galaxies whose central regions sport a "bar-shaped" feature composed of dust, gas, and stars.

SBa: tightly wound arms;
SBb: moderately wound arms;
SBc: loosely wound arms.

Irregular Galaxies are those shapeless (and usually small) objects similar to the Milky Way's Magellanic Clouds.

Ir: Irregular Galaxy.

Planetary Nebulae

The Vorontsov–Velyaminov (VV) Types system is the most common means of describing planetary nebulae, the remnants of dead solar mass stars. It is used in the famous PK catalog of planetaries.

1. Stellar;
2. Smooth disk (a: brighter center, b: uniform brightness, c: ring structure);
3. Irregular disk (a: irregular brightness distribution, b: ring structure);
4. Ring structure;
5. Irregular form (like a diffuse nebula);
6. Anomalous form (no structure).

The Dreyer (NGC) Codes

The Great NGC catalog, that old standby of both amateur and professional astronomers, features detailed descriptions for each of its objects. Unfortunately, these descriptions are expressed in a cryptic system of "codes." These codes are often modified with a large range of descriptive letters such as "p" for "pretty" as in "pretty bright." A complete list of these letters and their meanings can be found at http://www.seds.org/billa/ngc.html

- *Brightness:*

B	Bright
pB	Pretty bright
cB	Considerably bright
vB	Very bright
eB	Extremely bright
F	Faint
pF	Pretty faint
cF	Considerably faint
vF	Very faint
eF	Extremely faint

- *Size:*

L	Large
pL	Pretty large
cL	Considerably large
vL	Very large
eL	Extremely large
S	Small
pS	Pretty small
cS	Considerably small
eS	Extremely small
vS	Very small

- *Shape:*

R	Round
vlE	Very little extended
E	Elliptical
cE	Considerably extended
pmE	Pretty much extended
mE	Much extended
vmE	Very much extended
eE	Extremely extended

Object Abbreviations Used in This Book

dif = diffuse nebula

dst = double star

gal = galaxy

glb = globular star cluster

OC = open (galactic) star cluster

QSO = QUASAR

ref = reflection nebula

snr = supernova remnant

APPENDIX 4

The Urban Astronomer's Guide: Complete List of Objects

Object Name	Type	Constellation	Right Ascension	Declination	Magnitude	Size
M94	gal	CVn	12h51m08.7s	+41°05′16″	8.9	5.0′ × 3.5′
M51 (Whirlpool)	gal	CVn	13h30m07.0s	+47°09′49″	8.9	11.0′ × 7.0′
M106	gal	CVn	12h19m14.7s	+47°16′21″	8.3	17.4′ × 6.6′
M63 (Sunflower)	gal	CVn	13h16m04.6s	+41°59′53″	8.6	12.6′ × 7.5′
M81	gal	UMa	09h55m36.0s	+69°04′00″	6.9	21.0′ × 10.0′
M82	gal	UMa	09h56m22.0s	+69°39′27″	8.4	10.5′ × 5.1′
M101	gal	UMa	14h03m25.2s	+54°19′05″	7.9	22.0′ × 22.0′
M97 (Owl)	pln	UMa	11h15m06.5s	+54°59′22″	11.0	3.4′ × 3.3′
M 108	gal	UMa	11h11m51.9s	+55°38′47″	10.0	8.1′ × 2.7′
M3	glb	CVn	13h42m26.2s	+28°20′54″	6.3	18.0′
Cor Caroli	dst	CVn	12h56m17.5s	+38°17′13″	2.9	—
M65	gal	Leo	11h19m13.0s	+13°03′46″	10.2	9.1′ × 2.2′
M66	gal	Leo	11h20m32.3s	+12°57′36″	9.6	8.7′ × 4.1′
NGC 3628	gal	Leo	11h20m34.1s	+13°33′28″	10.3	12.0′ × 3.3′
NGC 2903	gal	Leo	09h32m28.3s	+21°28′42″	9.1	12.6′ × 5.5′
NGC 3190	gal	Leo	10h18m23.5s	+21°48′24″	12.1	5.5′ × 1.7′
NGC 3193	gal	Leo	10h18m43.1s	+21°52′04″	12.0	3.5′ × 1.8′
M105	gal	Leo	10h48m07.1s	+12°33′14″	9.6	4.8′
NGC 3384	gal	Leo	10h48m18.0s	+12°38′00″	10.0	5.9′
NGC 3389	gal	Leo	10h40m30.0s	+12°32′00″	12.0	2.7′
M95	gal	Leo	10h44m15.0s	+11°40′33″	9.7	6.0′ × 4.0′

(cont.)

Object Name	Type	Constellation	Right Ascension	Declination	Magnitude	Size
M96	gal	Leo	10h46m48.0s	+11°49'00"	9.3	6.0' × 4.0'
NGC 3521	gal	Leo	11h06m05.6s	−00°03'50"	9.7	9.5' × 5.4'
Algieba	dst	Leo	10h20m16.5s	+19°48'54"	2.2	—
Melotte 111	OC	Com	12h25m22.9s	+26°04'31"	2.9	5.0°
M64 (Black Eye)	gal	Com	12h57m00.0s	+21°39'09"	8.5	9.3' × 5.1'
NGC 4565	gal	Com	12h36m37.5s	+25°57'21"	9.6	14.8' × 2.1'
M53	glb	Com	13h13m11.0s	+18°08'22"	7.6	13.0'
NGC 5053	glb	Com	13h16m42.9s	+17°40'11"	9.8	9.0'
M88	gal	Com	12h32m15.9s	+14°23'22"	9.6	6.8' × 3.6'
M99	gal	Com	12h19m06.2s	+14°23'09"	9.9	5.3' × 4.6'
M100 (Catharine)	gal	Com	12h22m48.0s	+15°30'26"	12.6	4.7' × 1.1'
M85	gal	Com	12h25m40.8s	+18°09'36"	9.1	7.4' × 5.9'
NGC 4394	gal	Com	12h26m12.3s	+18°10'59"	10.9	3.5' × 3.3'
NGC 4725	gal	Com	12h50m42.9s	+25°28'10"	9.4	10.4' × 7.2'
NGC 4559	gal	Com	12h36m14.3s	+27°55'42"	10.5	11.0' × 4.9'
NGC 4147	gal	Com	12h10m22.8s	+18°30'39"	10.3	4.0'
24 Com	dst	Com	2h35m22.8s+	18°20'46"	6.7	—
M60	gal	Vir	12h43m56.6s	+11°31'19"	8.8	7.2' × 5.9'
NGC 4647	gal	Vir	12h43m49.0s	+11°33'05"	11.9	2.8' × 2.3'
M59	gal	Vir	12h42m18.7s	+11°37'02"	9.6	5.0' × 3.8'
M58	gal	Vir	12h38m00.5s	+11°47'16"	10.8	5.0' × 3.8'
M89	gal	Vir	12h35m42.0s	+12°33'00"	9.8	5.0' × 4.6'
M90	gal	Vir	12h37m06.6s	+13°07'55"	9.5	9.5' × 4.5'
M87	gal	Vir	12h31m05.7s	+12°21'38"	8.6	7.0'
NGC 4476	gal	Vir	12h30m15.7s	+12°19'05"	13.1	1.8' × 1.3'
M84	gal	Vir	12h25m20.6s	+12°51'21"	9.1	6.7' × 6.0'
M86	gal	Vir	12h26m28.8s	+12°54'54"	8.9	9.8' × 6.3'
NGC 4387	gal	Vir	12h25m58.4s	+12°46'45"	12.0	1.7' × 1.1'
NGC 4388	gal	Vir	12h26m03.5s	+12°37'53"	11.0	5.5' × 1.4'
Porrima	dst	Vir	12h41m56.0s	−01°28'46"	3.5	—
3C 273	QSO	Vir	12h29m23.3s	+02°01'19"	12.8	—
M12	glb	Oph	16h47m29.6s	−01°57'35"	6.1	16.0'
M10	glb	Oph	16h57m24.8s	−04°06'42"	6.6	20.0'
M5	glb	Ser	15h18m49.7s	+02°03'40"	5.6	22.0'
M 107	glb	Oph	16h32m48.9s	−13°04'01"	7.9	10.0'
NGC 6235	glb	Oph	16h53m24.0s	−22°11'00"	10.2	5.0'
M62	glb	Oph	17h01m32.0s	−30°07'19"	6.5	15.0'
NGC 6572	pln	Oph	18h12m06.0s	+06°51'00"	9.0	11.0"
Mirfak	dst	Her	16h08m18.8s	+17°02'08"	5.0	—
M80	glb	Sco	18h36m41.7s	−23°54'05"	5.1	32.0'
M17	dif	Sgr	18h21m03.9s	−16°09'38"	6.0	11.0'
M16	dif	Ser	18h19m06.6s	−13°45'48"	6.0	15.0'
M8 (Lagoon)	dif	Sgr	18h04m00.0s	−24°23'07"	5.0	45.0' × 30.0'
NGC 6530	OC	Sgr	18h05m03.9s	−24°18'52"	5.1	14.0'
M4	glb	Sco	16h23m54.5s	−26°32'17"	5.9	26.0'
M6	OC	Sco	17h40m06.0s	−32°13'00"	4.2	20.0'
NGC 6144	glb	Sco	16h27m32.4s	−26°02'04"	9.1	7.4'

Object Name	Type	Constellation	Right Ascension	Declination	Magnitude	Size
Antares	dst	Sco	16h29m42.9s	−26°26′41″	0.96	—
M39	OC	Cyg	21h32m20.9s	+48°27′25″	4.6	32.0′
M29	OC	Cyg	20h24m06.2s	+38°32′25″	7.5	6.0′
M71	glb?	Sge	19h53m58.3s	+18°47′11″	8.3	6.0′
NGC 6910	OC	Cyg	20h23m14.0s	+40°47′22″	7.3	7.0′
NGC 6866	OC	Cyg	20h03m52.6s	+44°00′03″	9.0	6.0′
NGC 6819	OC	Cyg	19h41m27.7s	+40°11′26″	9.5	5.0′
NGC 6834	OC	Cyg	19h52m23.4s	+29°25′14″	9.7	5.0′
NGC 6830	OC	Vul	19h51m14.4s	+23°04′10″	8.0	12.0′
NGC 6823	OC	Vul	19h43m19.6s	+23°18′36″	7.0	12.0′
Albireo	dst	Cyg	19h30m54.3s	+27°57′53″	3.1	—
M57 (Ring)	pln	Lyr	18h53m45.3s	+33°01′44″	9.4	1.1′
M27 (Dumbbell)	pln	Vul	19h59m48.1s	+22°43′47″	7.3	8.5′ × 5.7′
NGC 6826 (Blinking)	pln	Cyg	19h44m54.5s	+50°31′52	8.8	24.0″
NGC 6302 (Bug)	pln	Sco	17h14m04.2s	−37°06′38″	12.8	1.2′ × .5′
Cat's Eye	pln	Dra	17h58m32.1s	+66°37′30″	8.3	24.0″
Epsilon 1 Lyr	dst	Lyr	8h44m29.3s	+39°40′07″	5.0	—
Epsilon 2 Lyr	dst	Lyr	18h44m31.8s	+39°36′40″	5.2	—
M15	glb	Peg	21h30m11.2s	+12°11′06″	6.4	12.3′
M2	glb	Aqr	21h33m41.3s	−00°48′16″	6.5	8.0′
M56	glb	Lyr	19h16m46.7s	+30°11′17″	8.3	7.0′
NGC 6934	glb	Del	20h34m24.6s	+07°25′05″	8.9	5.9′
M72	glb	Aqr	20h53m43.5s	−12°31′15″	9.4	6.6′
Mesarthim	dst	Ari	01h53m47.4s	+19°19′08″	4.8	—
M31 (Andromeda)	gal	And	00h42m58.9s	+41°17′51″	4.0	178.0′ × 63.0′
NGC 206	OC	And	00h40m44.6s	+40°45′42″	4.0′	
M32	gal	And	00h42m56.5s	+40°53′39″	9.08	8.8′ × 6.5′
M110	gal	And	00h40m37.0s	+41°42′50″	8.93	21.9′ × 9.8′
M1 (Crab)	snr	Tau	05h34m48.5s	+22°01′20″	8.4	8.0′
NGC 404	gal	And	01h09m42.3s	+35°44′45″	11.3	3.4′
NGC 7331	gal	Peg	22h37m16.2s	+34°26′24″	10.3	10.7′ × 4.3′
Stephan's Quintet	gal	Peg	22h36m12.4s	+33°59′22″	12.0	3.2′
Almaak	dst	And	02h04m11.1s	+42°21′23″	2.3	—
E.T.	OC	Cas	01h19m19.9s	+58°21′30″	7.0	13.0′
NGC 436	OC	Cas	01h15m54.9s	+58°50′35″	8.8	6.0′
M103	OC	Cas	01h33m12.0s	+60°42′00″	7.4	6.0′
Trumpler 1	OC	Cas	01h35m57.7s	+61°19′03″	8.1	5.0′
NGC 654	OC	Cas	01h44m21.7s	+61°54′47″	6.5	5.0′
NGC 663	OC	Cas	01h46m21.2s	+61°16′43″	7.1	16.0′
IC 166	OC	Cas	01h52m50.1s	+61°51′30″	11.7	5.0′
M52	OC	Cas	23h24m23.9s	+61°37′08″	6.9	13.0′
NGC 7789	OC	Cas	23h57m14.2s	+56°45′24″	6.7	16.0′
NGC 129	OC	Cas	00h30m07.9s	+60°15′20″	6.5	21.0′

(cont.)

Object Name	Type	Constellation	Right Ascension	Declination	Magnitude	Size
NGC 189	OC	Cas	00h39m49.9s	+61°06′14″	11.1	3.7′
NGC 225	OC	Cas	00h43m40.8s	+61°49′12″	7.0	12.0′
NGC 133	OC	Cas	00h31m28.6s	+63°23′20″	9.4	7.0′
NGC 146	OC	Cas	00h33m17.7s	+63°19′19″	9.1	7.0′
King 14	OC	Cas	00h32m04.8s	+63°11′19″	8.5	7.0′
Achird	dst	Cas	00h49m21.3s	+57°50′38″	3.4	—
NGC 253 (Sculptor)	gal	Scl	00h47m47.0s	−25°15′49″	7.72	27.7′ × 6.8′
NGC 288	glb	Scl	00h52m58.8s	−26°33′32″	8.1	13.8′
M 77	gal	Cet	02h42m55.7s	+00°00′27″	9.64	7.0′ × 5.9′
NGC 1055	gal	Cet	02h42m00.1s	+00°27′46″	10.6	6.8′ × 3.2′
M30	glb	Cap	21h40m24.0s	+23°11′00″	7.5	11.0′
Kaffaljidhma	dst	Cet	02h43m33.1s	+03°15′24″	3.5	—
Merope Nebula	ref	Tau	03h46m29.5s	+23°47′03″	—	—
M78	ref	Ori	05h47m03.9s	+00°05′07″	8.0	8.0′ × 4.0′
NGC 2071	ref	Ori	05h47m21.9s	+00°18′07″	8.0	7.0′
NGC 2186	OC	Ori	06h12m26.8s	+05°27′10″	6.8	20.0′
NGC 2174	OC	Ori	06h09m42.6s	+20°40′04″	8.7	4.0′
NGC 2175	OC	Ori	06h09m54.6s	+20°29′03″	6.8	20.0′
NGC 2022	pln	Ori	05h42m23.1s	+09°05′22″	12.4	19.0″
NGC 1973	ref	Ori	05h35m21.3s	−04°43′50″	7.0	—
M42 (Orion)	dif	Ori	05h35m33.2s	−05°22′50″	4.0	90.0′
M43	dif	Ori	05h35m45.2s	−05°15′50″	9.0	20.0′
Meissa	dst	Ori	05h35m25.3s	+09°56′18″	3.6	—
M76	pln	Per	01h42m24.0s	+51°34′00″	11.0	3.0′ × 2.0′
M34	OC	Per	02h42m19.7s	+42°48′15″	5.2	35.0′
NGC 1023	gal	Per	02h40m41.8s	+39°05′14″	11.0	7.9′ × 3.5′
NGC 957	OC	Per	02h33m58.2s	+57°33′44″	7.6	11.0′
IC 2003	pln	Per	03h56m22.0s	+33°52′30″	12.0	8.0′
Miram	dst	Per	02h51m02.4s	+55°55′14″	3.8	—
M46	OC	Pup	07h42m02.6s	−14°48′52″	6.6	27.0′
NGC 2438	pln	Pup	07h42m05.4s	−14°44′53″	11.0	1.1′
M47	OC	Pup	07h36m50.8s	−14°29′29″	4.3	29.0′
M93	OC	Pup	07h44m51.0s	−23°52′06″	6.5	22.0′
NGC 2237 (Rosette)	dif	Mon	06h31m10.7s	+05°02′49″	5.5	80.0′ × 60.0′
NGC 2264 (Cone)	dif	Mon	06h41m17.4s	+09°53′46″	3.9	20.0′
NGC 2261 (Hubble's)	ref	Mon	06h39m29.2s	+08°44′46″	4.0	2.0′ × 1.7′
Adara	dst	CMa	06h58m50.2s	−28°58′51″	1.5	—
NGC 869 (h)	OC	Per	02h19m21.2s	+57°10′25″	4.3	29.0′
NGC 884 (Chi)	OC	Per	02h22m46.5s	+57°08′16″	4.4	29.0′
NGC 1528	OC	Per	04h15m46.5s	+51°15′31″	6.4	24.0′
NGC 2392 (Eskimo)	pln	Gem	07h29m29.7s	+20°54′09″	10.0	40.0″
NGC 1245	OC	Per	03h14m59.4s	+47°16′29″	8.4	10.0′
NGC 1545	OC	Per	04h21m14.7s	+50°16′07″	4.6	18.0′
Castor	dst	Gem	07h34m56.3s	+31°52′45″	1.9	—

Object Name	Type	Constellation	Right Ascension	Declination	Magnitude	Size
M 41	OC	CMa	06h47m16.7s	−20°43′44″	4.8	38.0′
NGC 2359 (Thor)	dif	CMa	07h18m36.0s	−13°12′00″	10.0′	—
Sirius	dst	CMa	06h45m22.9s	−16°43′27″	−1.5	—
NGC 2354	OC	CMa	07h14m28.5s	−25°43′52″	8.9	20.0′
NGC 2362	OC	CMa	07h18m59.9s	−24°56′12″	4.39	8.0′
NGC 2280	gal	CMa	06h45m01.2s	−27°38′45″	11.0	7.0′ × 3.0′

Index

Plate 1. Orion in light-polluted skies.

Plate 2. Celestron's 80 mm $f/5$ refractor.

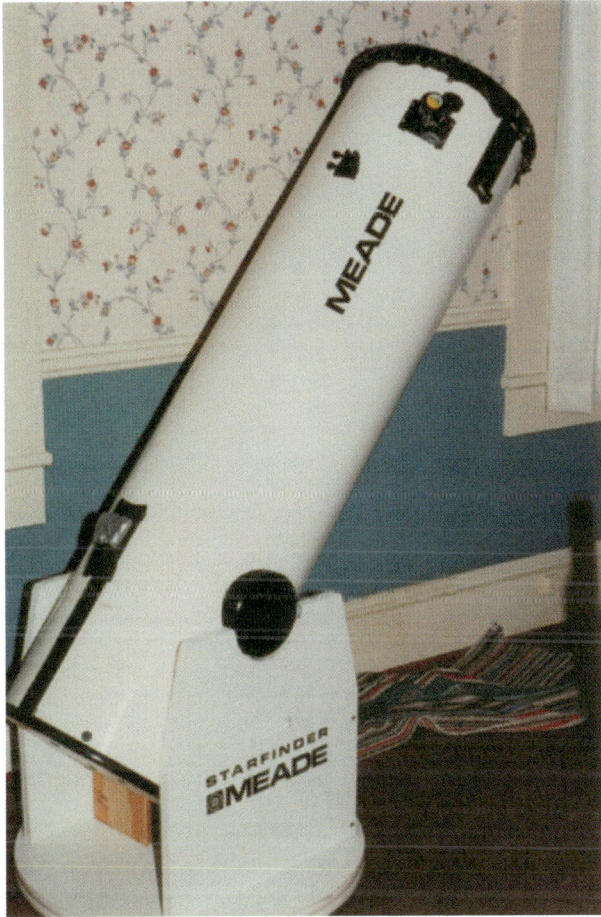

Plate 3. Meade's inexpensive but effective 12.5 inch Dobsonian.

Plate 4. A pair of truss tube style Dobsonian telescopes.

Plate 5. The ubiquitous Synta EQ4 German Equatorial Mount.

Plate 6. Celestron C8 Schmidt Cassegrain Telescope

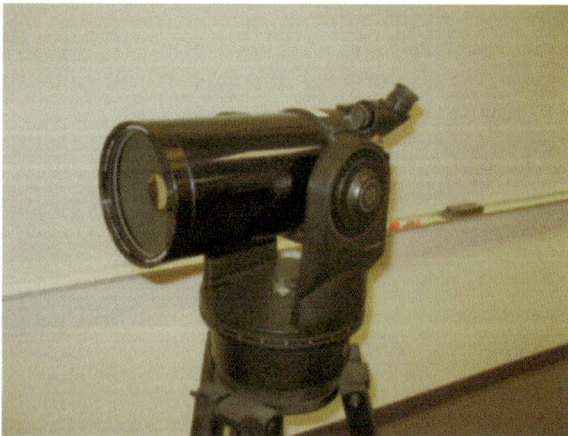

Plate 7. Meade ETX 125 Maksutov Cassegrain.

Plate 8. The beautiful Questar 3.5 Maksutov Cassegrain Telescope. (Courtesy of Jack Estes)

Plate 9. A selection of Light Pollution Reduction (LPR) filters.

Plate 10. Steve Kufeld's ingenious TELRAD aiming device.

Plate 11. Barlow lenses.

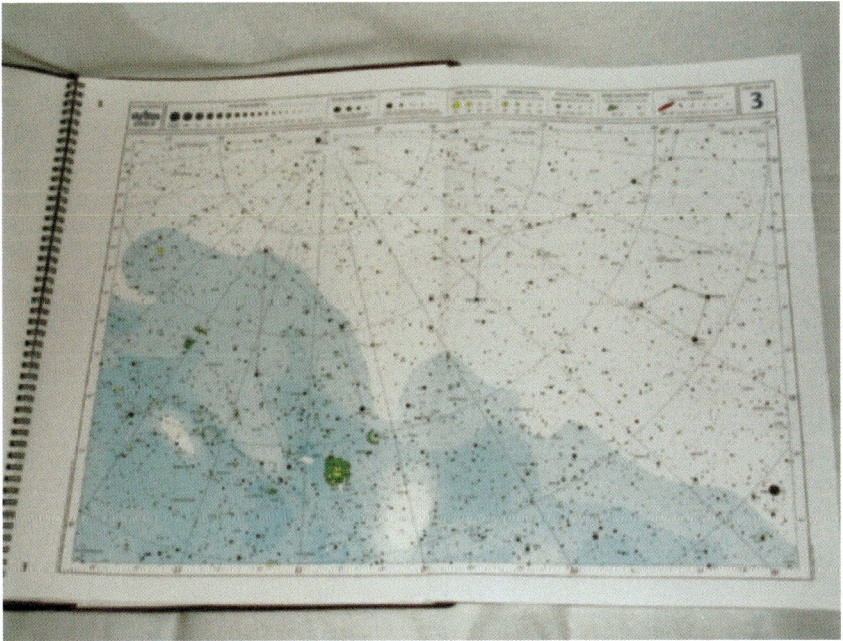

Plate 12. *Sky Atlas 2000.*

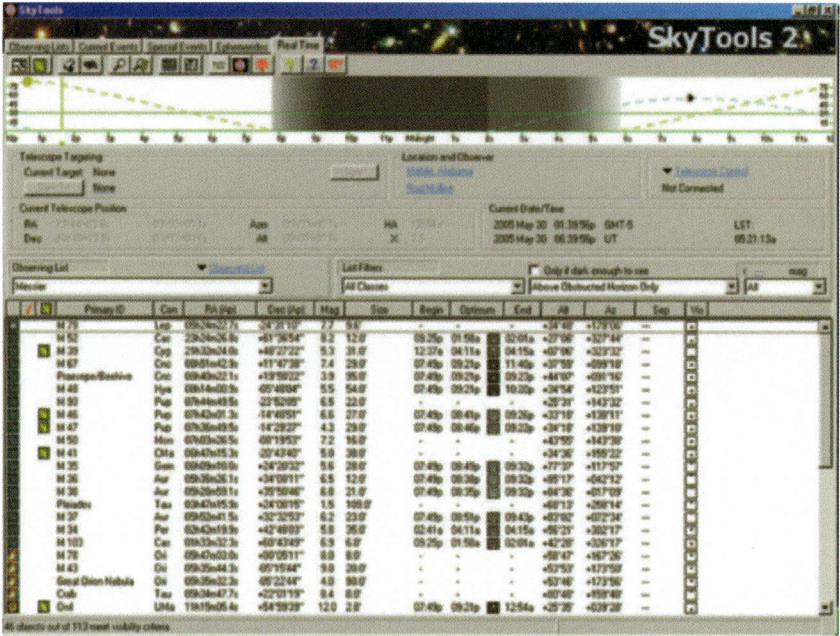

Plate 13. *Skytools 2* main display.

Plate 14. Stage flat light shield (unpainted to show detail).

Plate 15. A typical roll-off roof backyard observatory.

Plate 16. Night vision preserving red goggles.

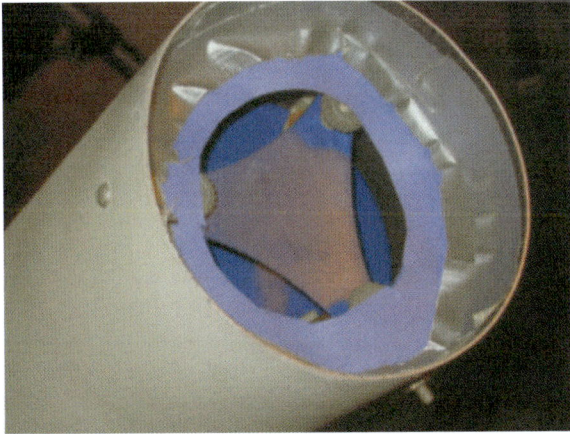

Plate 17. Simple mirror-end baffle for Newtonian.

Plate 18. The Denkmeier Deep Sky Binoviewer.

Plate 19. Old-fashioned analog setting circles.

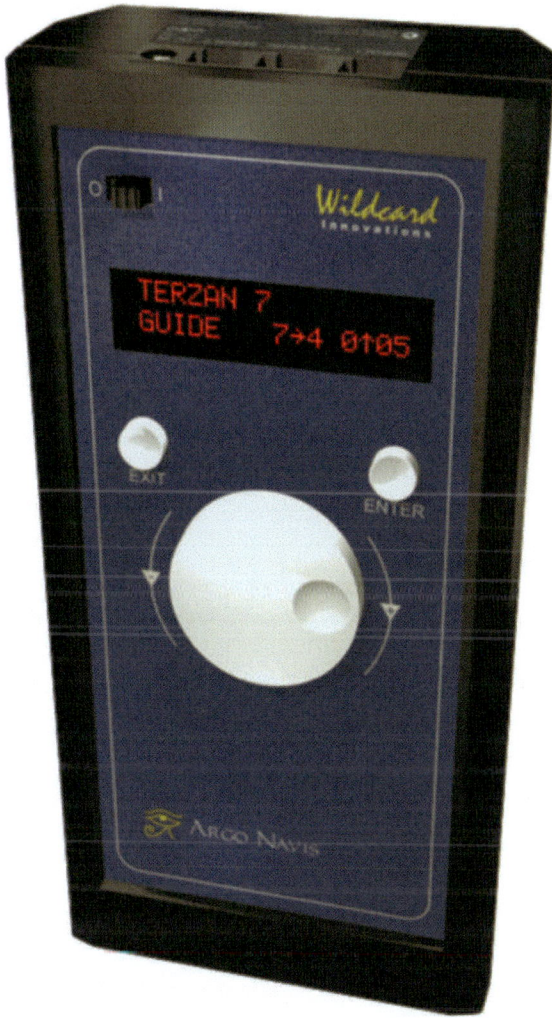

Plate 20. Argo Navis Digital Setting Circles. (Gary Kopff and Wildcard Innovations Inc.)

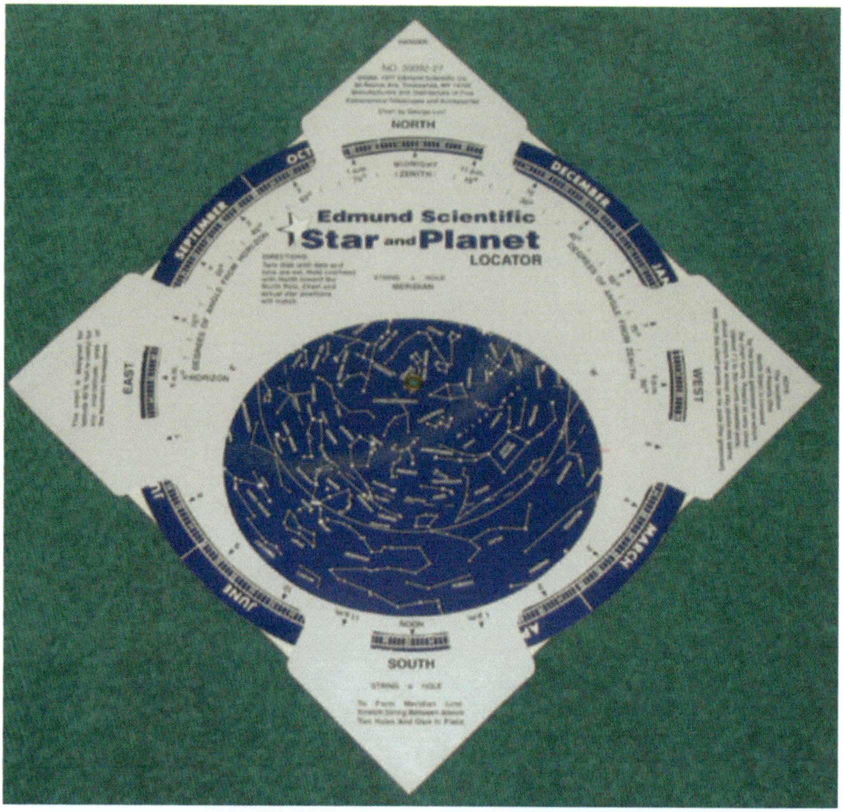

Plate 21. A planisphere type chart.

Plate 22. M42 at f/5 with 80 mm refractor.

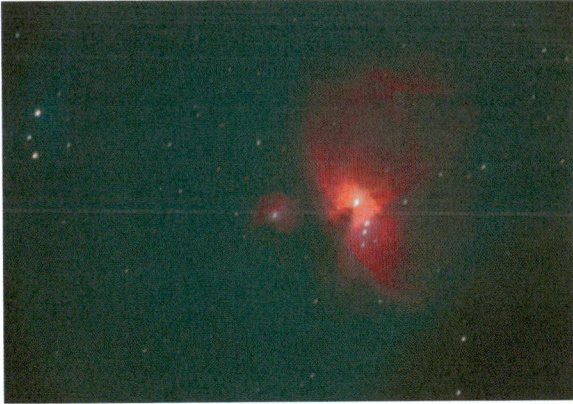

Plate 23. M42 at $f/10$ with C8 SCT.

Plate 24. M42 at $f/10$ with Fuji Super G 800 film.

Plate 25. Comet Hale Bopp piggyback image.

Plate 26. M13 image from heavily light polluted skies.

Plate 27. Video image of the Horsehead Nebula area (Courtesy of Jim Ferreira).

Plate 28. The City Lights Telescopes.

Plate 29. M51, The Whirlpool Galaxy, 8 inch SCT.

Plate 30. M64, The Blackeye Galaxy, 14 inch SCT.

Plate 31. (STScI)

Plate 32. Lonely M53. (Courtesy Space Telescope Science Institute (STScI) Digitized Sky Survey).

Plate 33. Mighty M87, 14 inch SCT.

Plate 34. M84 and M86. (STScl)

Plate 35. M10, 8 inch SCT.

Plate 36. Serpens' splendid M5, 8 inch SCT.

Plate 37. M17, The Swan, swims on. (STScI)

Plate 38. M16, The Eagle. (STScI)

Plate 39. Difficult little M71 with an 8 inch SCT.

Plate 40. The Dumbbell imaged from city lights.

Plate 41. The Bug Nebula. (STScI)

Plate 42. Cosmic Cat's Eye. (STScI)

Plate 43. M56 imaged by an 8 inch SCT and CCD cam.

Plate 44. Urban Andromeda astrophoto.

Plate 45. A beautiful Palomar print of M1 (STScI)

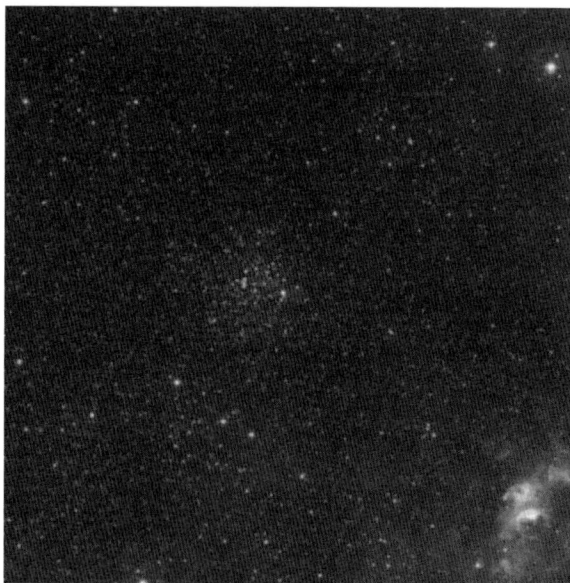

Plate 46. M52 and the faint Bubble Nebula (lower right). (STScI)

Plate 47. NGC 7789. (STScI)

Plate 48. Enormous and detailed galaxy NGC 253. (STScI)

Plate 49. Amateur CCD image of peculiar galaxy M77.

Plate 50. The field of M78 is filled with reflection nebulosity. (STScI)

Plate 51. The Little Dumbbell with 8 inch scope.

Plate 52. NGC 1023 in a big scope under dark skies. (STScI)

Plate 53. Planetary nebula NGC 2438. (STScI)

Plate 54. A portion of the giant Rosette Nebula. (STScI)

Plate 55. Winter's Eskimo. (STScI)

Plate 56. Thor's horned helmet, NGC 2359. (STScI)

Plate 57. NGC 2280, galaxy in Canis Major. (STScI)

21705113R00166

Printed in Great Britain
by Amazon